Berliner ökophysiologische

und phytomedizinische Schriften

Hrsg. von Christian Ulrichs und Carmen Büttner

Lebenswissenschaftliche Fakultät,

Humboldt-Universität zu Berlin

Band 50

Suitability of portable NIR sensors (food-scanners) for the determination of fruit quality along the supply chain using the example of tomatoes

D I S S E R T A T I O N
zur Erlangung des akademischen Grades

Doctor rerum horticulturarum
(Dr. rer. hort.)

eingereicht an der
Lebenswissenschaftlichen Fakultät der Humboldt-Universität zu Berlin

von

M. Sc. Simon Goisser

Präsidentin
der Humboldt-Universität zu Berlin

Prof. Dr.-Ing. Dr. Sabine Kunst

Dekan der Lebenswissenschaftlichen Fakultät
der Humboldt-Universität zu Berlin

Prof. Dr. rer. nat. Dr. rer. agr. Christian Ulrichs

Gutachter/innen

1. Prof. Dr. rer. nat. Dr. rer. agr. Christian Ulrichs
2. Dr. rer. habil. Manuela Zude-Sasse

Tag der mündlichen Prüfung: 25.10.2021

Bibliografische Information der Deutschen Nationalbibliothek
Die Deutsche Nationalbibliothek verzeichnet diese Publikation in der Deutschen Nationalbibliografie; detaillierte bibliographische Daten sind im Internet über http://dnb.d-nb.de abrufbar.
1. Aufl. - Göttingen: Cuvillier, 2021
Zugl.: Berlin, Univ., Diss., 2021

Die vorliegende Arbeit wurde als Dissertation an der Lebenswissenschaftlichen Fakultät der Humboldt-Universität zu Berlin eingereicht und am 25.10.2021 verteidigt.

Nonnenstieg 8, 37075 Göttingen
Telefon: 0551-54724-0
Telefax: 0551-54724-21
www.cuvillier.de

1. Auflage, 2021
Gedruckt auf umweltfreundlichem, säurefreiem Papier aus nachhaltiger Forstwirtschaft.

ISBN 978-3-7369-7543-9
eISBN 978-3-7369-6543-0

Preface

The content of this thesis is based on reformatted work, which has been published as follows:

1. Goisser, S.; Fernandes, M.; Ulrichs, C.; Mempel, H. (2018): Non-destructive measurement method for a fast quality evaluation of fruit and vegetables by using food-scanner. DGG-Proceedings 8 (13), p. 1-5. DOI: https://doi.org/10.5288/dgg-pr-sg-2018.

2. Goisser, S.; Krause, J.; Fernandes, M.; Mempel, H. (2019): Determination of tomato quality attributes using portable NIR-sensors. In: Jürgen Beyerer, Fernando Puente León und Thomas Längle (Hg.): OCM 2019 - 4th International Conference on Optical Characterization of Materials. March 13th - 14th, 2019. Karlsruhe: Scientific Publishing, p. 1–12. DOI: https://doi.org/10.5445/KSP/1000087509.

3. Goisser, S.; Mempel, H.; Bitsch, V. (2020): Food-Scanners as a Radical Innovation in German Fresh Produce Supply Chains. International Journal on Food System Dynamics 11 (2), p. 101-116. DOI: https://doi.org/10.18461/ijfsd.v11i2.43.

4. Goisser, S.; Wittmann, S.; Fernandes, M.; Mempel, H.; Ulrichs, C. (2020): Comparison of colorimeter and different portable food-scanners for non-destructive prediction of lycopene content in tomato fruit. Postharvest Biology and Technology 167 (111232). DOI: https://doi.org/10.1016/j.postharvbio.2020.111232.

5. Goisser, S.; Fernandes, M.; Wittmann, S.; Ulrichs, C.; Mempel, H. (2020): Evaluating the practicability of commercial food-scanners for non-destructive quality assessment of tomato fruit. Journal of Applied Botany and Food Quality 93, p. 204–214. DOI: https://doi.org/10.5073/JABFQ.2020.093.025.

6. Goisser, S.; Wittmann, S.; Mempel, H. (2021): Food-scanner applications in the fruit and vegetable sector. Landtechnik 76 (1), p. 52-76. DOI: https://doi.org/10.15150/lt.2021.3264.

Summary

The quality of fruit and vegetables is subject to continuous postharvest change due to ongoing metabolic and post-ripening processes. In the course of the supply chain from production to the end-consumer various factors, e.g., storage and transport conditions, mechanical stress and packaging, determine the degree and speed of further degeneration of product quality. The task of postharvest quality assurance is to monitor the quality of fresh produce and to counteract processes that reduce quality in order to keep losses along the value chain as low as possible. Optical measuring methods, such as visual and near-infrared spectroscopy (VIS/NIRS), offer numerous advantages over conventional measuring methods for determining important quality parameters of fruit and vegetables, such as fast and non-destructive measurements and simple sample preparation. The need for a more cost-effective and compact application of NIR spectroscopy in contrast to conventional laboratory instruments has led to the development of a wide range of portable handheld devices and miniaturized sensors in recent years. In contrast to laboratory instruments, these devices allow on-site and in-field measurements. In the case of fruit and vegetables, quality measurements can be carried out at various points along the entire supply chain, therefore the actual quality can be determined non-destructively. Some commercial portable and miniaturized instruments have been developed specifically for fresh produce and are designed to help the user to determine key quality parameters in the orchard, in the incoming goods control or at the point of sale in retail outlets.

The present study evaluated the suitability of these portable and partly miniaturized NIR sensors, so-called food-scanners, for determining fruit quality along the supply chain of fruit and vegetables. Using the approach of qualitative research, the first step comprised interviews of experts at different positions of the fruit and vegetable value chain. In this way preferences and concerns regarding the implementation of this technology for fresh produce were investigated. Based on the findings, non-destructive prediction models for a number of important quality characteristics and secondary plant constituents were developed on the model fruit tomato. In addition, the suitability of the practical use of these devices was evaluated and a first trial for use in daily incoming goods control was carried out.

The results of the qualitative study showed a wide variation in the practices currently used in quality control to determine product quality, particularly between production companies and wholesale and retail businesses. In general, the determination of quality often depends to a large extent on trained and experienced personnel who test the products subjectively and oftentimes use haptical and visual inspections. In addition, wholesalers often follow established protocols to ensure quality and conformity with the given statuatory standards. The results show that food-scanners can facilitate quality control at different levels of the fresh produce value chain by providing fast, non-destructive and objective measurements, while helping to overcome existing discrepancies in quality assessment along the supply chain by introducing a common measurement method. Concerns about the use of food-scanners have been identified with regard to possible additional requirements within the fruit trade, which could lead to more pressure and additional obligations for producers.

Subsequent studies evaluated the performance of three commercially available portable and miniaturized VIS/NIR spectrometers. The studies focused on the non-destructive prediction of various important quality characteristics of the model fruit tomato. The evaluation of the food-scanner spectra led to prediction models with high correlations ($r^2 > 0.90$) for the quality traits firmness, dry matter, sugar content and the colour values L*, a* and h°. In the course of further investigations, prediction models with high prediction capabilities ($r^2 > 0.92$) were obtained for the non-destructive determination of the secondary plant metabolite lycopene in intact tomatoes. In addition, the software provided by the sensors manufacturers to create prediction models was evaluated by comparing the results derived from these software tools with state-of-the-art software for multivariate analysis. The direct comparison of the evaluation software showed very similar accuracies for the calculated prediction models. In a further experiment, the decrease in firmness, which is considered an important indicator for the shelf life of tomatoes, could be modelled over the storage period in connection with the respective storage temperature. By combining these firmness models with non-destructive NIR predictions of the current firmness, a first approach for a shelf life model of tomatoes could be developed. Investigations comparing the performance of a food-scanner with a conventional laboratory spectrometer for the prediction of the tomato quality parameters sugar content and firmness underlined the high accuracy of the food-scanner predictions and the comparability to laboratory instruments.

Based on the results derived from tomatoes, further studies were carried out to predict important quality parameters in a wide range of fruit and vegetables (apple, avocado, blueberry, ginger, persimmon, kaki, kiwi, mandarin, mango, table grapes). The evaluations showed that in particular the quality traits sugar content, dry matter as well as the relative water content of fruit can be predicted non-destructively with high accuracy. In a concrete practical example, a food-scanner was also used for non-destructive quality assessment in the course of incoming goods control of a fruit and vegetable wholesaler. The evaluation showed a high degree of agreement between the food-scanner predictions and conventional destructive measurements using refractometers and furthermore illustrated the possibility of saving working time by using these measuring devices.

The results of this doctoral thesis illustrate the great potential of food-scanners for the non-destructive quality measurement of fruit and vegetables along the fresh produce supply chain. Some of the commercially available devices investigated in these studies already showed a high degree of compatibility with the requirements of the fruit and vegetable industry, such as a robust design, long battery life and the possibility of personalized modelling and data evaluation by users. In addition to the possibility of integrating the generated data into existing information technology systems, confidence in the accuracy of the food-scanner predictions was identified as an important prerequisite for implementation in practice. Targeted cooperation between food-scanner manufacturers and companies in the fruit and vegetable sector, the publication of new findings in specialized journals for producers and trading companies and the provision of targeted training programmes could lead to a better adaption of food-scanner to the requirements of the fresh produce supply chain in the future. Ultimately, this could allow the implementation of these novel devices in everyday practice of fruit quality control.

Zusammenfassung

Die Qualität von Obst und Gemüse unterliegt nach der Ernte einer kontinuierlichen Veränderung auf Grund anhaltender Stoffwechsel- und Nachreifeprozesse. Im Verlauf der Wertschöpfungskette von der Produktion bis zum Endkunden bestimmen zudem verschiedene Einflussfaktoren, wie beispielsweise Lager- und Transportbedingungen, mechanische Belastung sowie Verpackung, den Grad sowie die Geschwindigkeit der weiteren Degeneration der Produktqualität. Die Aufgabe der Qualitätssicherung nach der Ernte ist dabei, die Qualität der Frischeprodukte zu überwachen und qualitätsmindernden Prozessen entgegenzuwirken, um Verluste entlang der Wertschöpfungskette so gering wie möglch zu halten. Optische Messmethoden, wie beispielsweise die Visuelle- und Nahinfrarot-Spektroskopie (VIS/NIRS), bieten gegenüber herkömmlichen Messverfahren zur Bestimmung wichtiger Qualitätsparameter an Obst und Gemüse zahlreiche Vorteile, wie beispielsweise eine schnelle und zerstörungsfreie Messung sowie eine einfache Probenvorbereitung. Der Bedarf nach einer kostengünstigeren und kompakteren Anwendung der NIR-Spektroskopie im Gegensatz zu den herkömmlichen Laborinstrumenten führte in den vergangenen Jahren zur Entwicklung verschiedenster tragbarer Handgeräte sowie miniaturisierter Sensoren. Im Gegensatz zu Laborinstrumenten ermöglichen diese Geräte eine Messung vor Ort. Im Fall von Obst und Gemüse können Qualitätsmessungen an verschiedenen Stellen der gesamten Wertschöpfungskette durchgeführt und so tatsächliche Produktqualitäten zerstörungsfrei ermittelt werden. Einige kommerzielle portable und miniaturisierte Geräte wurden speziell für Frischeprodukte entwickelt und sollen dem Anwender dabei helfen, wichtige Qualitätsparameter in der Obstplantage, der Warenkontrolle oder am Verkaufsort im Einzelhandel zu bestimmen.

In der vorliegenden Studie wurde die Eignung dieser tragbaren und zum Teil miniaturisierten NIR-Sensoren, sogenannten Food-Scannern, für die Bestimmung der Fruchtqualität entlang der Wertschöpfungskette von Obst und Gemüse evaluiert. Mittels Methoden der qualitativen Sozialforschung wurden in einem ersten Schritt Experten an verschiedenen Positionen der Obst und Gemüse Wertschöpfungskette befragt und Präferenzen und Bedenken hinsichtlich der Implementierung dieser Technologie für Frischeprodukte untersucht. Aufbauend auf den gewonnenen Erkenntnissen wurde an der Modellfrucht Tomate zerstörungsfreie Vorhersagemodelle für eine Vielzahl wichtiger Qualitätsmerkmale sowie sekundärer Pflanzeninhaltsstoffe entwickelt. Darüber hinaus wurde die Eignung des praktischen Einsatzes dieser Geräte evaluiert und ein erster Versuch zur Verwendung in der täglichen Wareneingangskontrolle durchgeführt.

Die Ergebnisse der qualitativen Befragung zeigten eine große Variation der aktuell in der Qualitätskontrolle durchgeführten Praktiken zur Ermittlung der Produktqualität, insbesondere zwischen den Unternehmen der Produktion und des Groß- und Einzelhandels. Generell hängt die Qualitätsbestimmung oft in hohem Maße von geschultem und erfahrenem Personal ab, welches die Produkte subjektiv und meist haptisch und visuell prüft. Darüber hinaus halten sich Großhändler oft an etablierte Protokolle, um Qualität und Konformität mit den gegebenen rechtlichen Standards zu gewährleisten. Die Ergebnisse verdeutlichen, dass Food-Scanner die Qualitätskontrolle auf den verschiedene Ebenen der Wertschöpfungskette von Frischeprodukten durch schnelle, zerstörungsfreie und objektive Messungen erleichtern und gleichzeitig dabei helfen können, bestehede Diskrepanzen bei der Qualitätsbeurteilung entlang der Lieferkette durch die Einführung einer einheitlichen Messmethode zu überwinden. Bedenken über den Einsatz von Food-Scannern konnten im Hinblick auf mögliche zusätzliche Anforderungen des Fruchthandels identifiziert werden, die zu mehr Druck und zusätzlichen Auflagen für die Produzenten führen könnten.

In den anschließenden Untersuchungen wurde die Leistungsfähigkeit von drei kommerziell erhältlichen portablen und miniaturisierten VIS/NIR-Spektrometern bewertet. Im Fokus der Studien stand die zerstörungsfreie Vorhersage verschiedener wichtiger Qualitätsmerkmale der Modellfrucht Tomate. Die Auswertung der Food-Scanner Spektren führte zu Vorhersagemodellen mit hohen Korrelationen ($r^2 > 0,90$) für die Merkmale Festigkeit, Trockenmasse, Zuckergehalt sowie die Farbwerte L*, a* und h°. Im Zuge weiterer Untersuchungen konnten zudem Vorhersagemodelle mit hohen Vorhersagegüten ($r^2 > 0,92$) für die zerstörungsfreie Bestimmung des sekundären Pflanzeninhaltsstoffs Lycopin in Tomaten erzielt werden. Zusätzlich wurde die von den Herstellern der Sensoren zur Verfügung gestellte Software zur Erstellung von Vorhersagemodellen evaluiert, indem die von diesen Software-Tools abgeleiteten Ergebnisse mit modernster Software für multivariate Analysen verglichen wurden. Der direkte Vergleich der Auswertungs-Software ergab sehr ähnliche Genauigkeiten für die berechneten Vorhersagemodelle. In einem weiteren Versuch konnte die Abnahme der Festigkeit, welche als wichtiger Indikator für die Haltbarkeit von Tomaten gilt, über die Lagerzeit in Zusammenhang mit der jeweiligen Lagertemperatur modelliert werden. Durch die Kombination dieser Festigkeitsverlaufsmodelle mit zerstörungsfreien NIR-Vorhersagen zur aktuellen Festigkeit konnte ein erster Ansatz für ein Haltbarkeitsmodell von Tomaten entwickelt werden. Untersuchungen, welche die Leistungsfähigkeit eines Food-Scanners mit einem herkömmlichen Labor-Spektrometer für die Vorhersage der Tomatenqualtätsparameter Zuckergehalt und Festigkeit verglichen, verdeutlichten die hohe Genauigkeit der Food-Scanner Vorhersagen und die Vergleichbarkeit zu Labormessungen.

Basierend auf den an Tomaten gewonnenen Erkenntnissen wurden weitere Untersuchungen zur Vorhersage wichtiger Qualitätsparameter an zahlreichen Obst- und Gemüsearten (Apfel, Avocado, Blaubeere, Ingwer, Kaki, Kiwi, Mandarinen, Mango, Tafeltrauben) durchgeführt. Die Auswertungen zeigten, dass insbesondere die Merkmale Zuckergehalt, Trockenmasse sowie der relative Wassergehalt der Früchte mit hoher Genauigkeit zerstörungsfrei vorhergesagt werden können. In einem konkreten Praxisbeispiel wurde zudem ein Food-Scanner im Zuge von Wareneingangskontrollen des Obst- und Gemüsegroßhandels zur zerstörungsfreien Qualitätsbewertung eingesetzt. Die Evaluierung ergab eine hohe Übereinstimmung der Food-Scanner Vorhersagen mit herkömmlichen zerstörerischen Messungen mittels Refraktometer und veranschaulichte zudem die Möglichkeit der Einsparung von Arbeitszeit durch den Einsatz dieser Messgeräte.

Die Ergebnisse dieser Doktorarbeit veranschaulichen das große Potential von Food-Scannern für die zerstörungsfreie Qualitätsmessung von Obst und Gemüse entlang der Wertschöpfungskette. Einige der untersuchten kommerziellen Geräte zeigten bereits eine hohe Kompatibilität mit den Anforderungen der Obst- und Gemüsebranche, wie beispielsweise eine robuste Bauweise, lange Batterielaufzeit sowie die Möglichkeit der personalisierten Modellbildung und Datenauswertung durch die Anwender. Als wichtige Voraussetzungen für die Implementierung in der Praxis wurden neben der Möglichkeit der Integrierung der generierten Daten in bestehende Systeme der Informationstechnologie auch das Vertrauen in die Genauigkeit der Food-Scanner-Vorhersagen ermittelt. Eine gezielte Zusammenarbeit der Food-Scanner Produzenten mit den Unternehmen der Obst- und Gemüsebranche, die Publikation von neuen Erkenntnissen in Fachzeitschriften für Produzenten und Handelsunternehmen sowie die Bereitstellung von gezielten Weiterbildungsprogrammen könnte dazu führen, dass Food-Scanner in Zukunft noch besser auf die Anforderungen der Wertschöpfungskette von Frischeprodukten abgestimmt werden und letztendlich die Implementierung in den praktischen Alltag der Qualitätskontrolle finden.

Acknowledgement

This doctoral dissertation is the result of experimental research conducted at the Univerity of Applied Sciences Weihenstephan-Triesdorf (HSWT) in collaboration with the Division Urban Plant Ecophysiology of the Faculty of Life Sciences, Humboldt Universität zu Berlin. This thesis was financially supported by the Bavarian Ministry of Food, Agriculture and Forestry as part of the alliance "Wir retten Lebensmittel" [We Save Foodstuffs] from June 2017 to August 2019 as well as the QS Science Funds in Fruit, Vegetables and Potatoes within the research project "Non-destructive quality assessment of fruit and vegetables along the supply chain using food-scanners" from October 2019 to September 2020. STEP Systems GmbH from Nuremberg provided two of the food-scanners used for this research, for which I am extremely thankful.

First and foremost I would like to thank my supervisor Prof. Dr. Heike Mempel and my thesis supervisor Prof. Dr. Dr. Christian Ulrichs for giving me the opportunity to conduct this research and enabling this thesis. I'm very grateful for their guidance during the course of this work. In particular, I would like to thank Prof. Dr. Heike Mempel for arranging the numerous contacts within the fruit and vegetable sector, for her consistent motivation, supervision, feedback and insight into practice, and for her input on how to design clear and concise presentations. I also want to acknowledge Prof. Dr. Vera Bitsch, chair of Economics of Horticulture and Landscaping at the Technical University of Munich, for her guidance and support during planning and realization of the qualitative study.

I am extremely thankful for the support of my team at Weihenstephan, especially my workmate Sabine Wittmann for her constant feedback, constructive criticism and mental support whenever I needed it, and for her and Ivonne's helping hands during the many experiments conducted within my three years work. A big thank you to Dietmar, Robert, Mica, Timo, Alexander, Karin, Bartl, Katrin, Torsten, Elke, Dominikus, Dr. Lohr and the soil science laboratory team, Prof. Dr. Jakob and his laboratory team, the guys from the workshop – you all have been a great source of support. I am also thankful to the colleagues at our cooperating institutions within the Ministry's research project, in particular to Michael Fernandes from the Deggendorf Institute of Technology, who got not tired in supporting me with multivariate evaluations. I also want to express my thanks to the staff at the Greenhouse Laboratory Centre in Dürnast, who were a big help during my lycopene analysis, and for Susanne Steger's assistance in gaining the extracurricular credits for my dissertation.

Special thanks go to my family, who supported and believed in me, and to my partner Natalie for her continuous motivation during the time of this dissertation.

Index

Alphabetical list of abbreviations and symbols

Abbreviation / symbol	Description
a*	Color value indicating the red/green coordinate
b*	Color value indicating the yellow/blue coordinate
C*	Color value indicating the chroma
CA	Cluster analysis
DA	Discriminant analysis
FEFO	First-expired-first-out
FIFO	First-in-first-out
FSC	Fresh produce supply chain
h°	Color value indicating the hue
KNN	K-nearest neighbor
L*	Color value indicating the lightness
LDA	Linear discriminant analysis
LED	Light-emitting diode
MAA	Multi-Actor Approach
MNPLS	Mixed-norm partial least squares
NIPALS	Nonlinear iterative partial least squares
NIRS	Near infrared spectroscopy
PCA	Principal component analysis
PLSR	Partial least squares regression
r^2_C	Coefficient of determination in calibration
r^2_{CV}	Coefficient of determination in cross validation
$RMSE_C$	Root mean square error of calibration
$RMSE_{CV}$	Root mean square error of cross validation
SIMCA	Soft independent modeling of class analogy
SPLS	Sparse partial least squares
SSC	Soluble sugar content
SVM	Support vector machine
VIS	Visual spectrum

List of figures

List of tables

1 Introduction

1.1 Problem statement and research questions

For the field of quality assurance, a comprehensive and holistic approach from production to marketing is of high importance. Within the fruit and vegetable sector, this is particularly important, as various factors during the production of fresh produce, such as mineral nutrition, climatic conditions, selection of variety, or maturity at harvest have a significant influence on the subsequent quality along the supply chain (Hewett 2006). Furthermore, the quality of fruit and vegetables changes immediately after harvest due to ongoing metabolic processes (Martínez-Romero et al. 2007). The quality of produce deteriorates at a different pace (Kong and Singh 2016) depending on various influencing factors during the further course along the supply chain (e.g., storage and transport conditions, packaging, mechanical handling, post-ripening processes). Postharvest quality assurance has to monitor the quality of fresh produce and counteract quality-reducing processes in order to keep potential losses along the supply chain as low as possible. Depending on the respective product, specific standards concerning the commercial quality of some fruit parameters must be met to allow the distribution via retail chains (UNECE 2019). In addition to these standards required by law, retail companies oftentimes impose requirements, which exceed statutory provisions.

The verification and assessment of quality of intact fruit can be made based on human senses, such as visual inspection of colors, smelling of volatile organic compounds or touching to control fruit firmness (Walsh 2006). However, these judgements are often subjective, therefore the quality of these evaluations highly depends on personal experience. The utilization of instrumental measurement methods for commercially important quality traits is often preferred over sensory evaluations and help to reduce the variation between individuals, provide high precision and offer a common language between consumers, researchers and industry (Abbott 1999). Furthermore, the determination of internal quality standards is often simply not possible using solely sensory assessments. In order to measure and verify these internal quality parameters, destructive measurements have to be conducted, e.g. the determination of sugar content using pressed fruit juice and a refractometer or measurement of firmness via penetrometer. Other measurements, such as the determination of dry matter, are extremely time-consuming, and in the case of acidity measurement, elaborate sample preparation and handling of chemicals is necessary for titration (OECD 2018).

In the past, various studies investigated the feasibility of optical measurement methods like visible and near-infrared (VIS/NIR) spectroscopy for the determination of important quality parameters of various agricultural and horticultural products (Nicolaï et al. 2007; Cortés et al. 2019). NIR spectroscopy, as a type of vibrational spectroscopy, employs the stimulation of molecular vibration using infrared light in the wavelength range 750 to 2500 nm. NIR electromagnetic waves interact with various molecular bonds (especially C-H, O-H, N-H or S-H bonds) of the respective sample constituents and result in a spectra, which is recorded by an optical sensor (Pasquini 2003). This NIR spectral data can be evaluated using various algorithms and multivariate statistical analysis and allows the development of qualitative and quantitative prediction models (Cozzolino 2009; Pasquini 2018). Whereas qualitative models are designed to enable classification, identification and authenticity verification of products (e.g., origin, variety, grade), quantitative models aim to determine the concentration of known constituents in samples, such as sugar concentration or dry matter content of fruit and vegetables (Figure 1). NIR spectroscopy offers various advantages compared to traditional measurement methods of fruit and vegetables, such as a fast and non-destructive operating principle, simple sample preparation as well as the possibility of using one spectrum for multidimensional evaluation of various parameters in one work step (McClure 1994; OECD 2018).

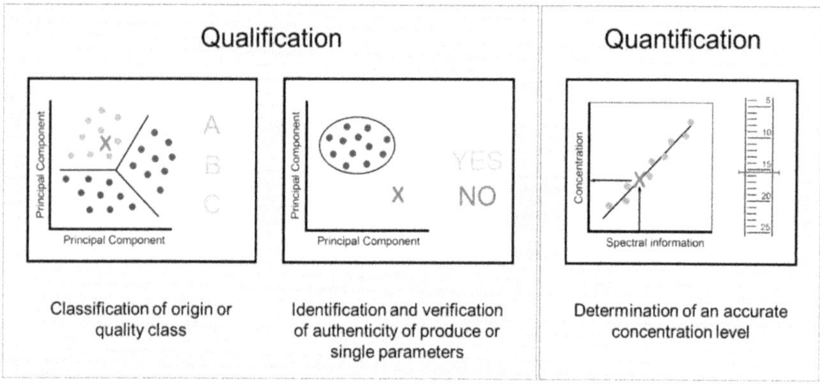

Figure 1: Application types of NIR spectroscopy (adapted and redesigned according to Langer (2019))

Next to laboratory applications, which employ benchtop instruments, the technique of NIR spectroscopy found its way into automated sorting and grading machines for fruit and vegetables. These on-line NIR applications are designed for real-time determination of important quality attributes (e.g., sugar content, acidity) and the detection of internal fruit disorders such as internal browning of fruit (Huang et al. 2008). The need for a more cost-effective and compact application of NIR spectroscopy led to the development of portable handheld devices (dos Santos et al. 2013). These sensors allow the transfer of traditional laboratory work to in-field applications, which can be implemented along the whole supply chain of fresh produce. Some companies are already marketing devices specifically designed for horticultural applications, such as the determination of fruit quality parameters in the orchard (Felix Instruments 2020b) as well as measurements along the supply chain to the point of sale (Sunforest 2020).

Subsequent technological developments led to an increasing miniaturization of NIR sensors. Some of these devices, so called food-scanners, were initially designed for end-consumers and were commercially distributed by various start-up companies (Consumer Physcis 2020; Tellspec Inc. 2020a; Spectral Engines Oy 2020). Oftentimes, these food-scanners work in combination with a mobile app on a smartphone or tablet and make use of cloud-based platforms, which hold vast material libraries and employ advanced algorithms. These applications aim to identify various food-related parameters (e.g. protein, fat, calories, allergens, contaminants, macronutrients) as well as the total energy content and provide important information such as food adulteration, food fraud, and food quality (Rateni et al. 2017). Currently, some of these start-up companies collaborate with industrial partners and target real-time in-field solutions for specific industries, e.g., monitoring of corn, grains and animal feed quality (Consumer Physcis 2020) as well as the analysis of fish, fruit and liquids (Tellspec Inc. 2020a).

Due to the novelty of portable and miniaturized devices, scientific research about the performance and suitability of food-scanners for a reliable determination of the quality of fruit and vegetables is currently still sparse. Initial studies compared various commercially available instruments for their prediction accuracy of specific fruit quality traits such as dry matter (Kaur et al. 2017) or examined the performance of individual devices (Ncama et al. 2018; Li et al. 2018; Choi et al. 2017) . Most of these studies regarding the performance of food-scanners evaluated selected fruit within a laboratory setup without reference to the practical use of these devices in daily processes of quality control along the fresh produce supply chain (FSC). At the beginning of this work, the use of miniaturized sensors for the concrete determination of fruit quality and their application along the FSC had not yet been investigated. In this context, the overall aim of this study was to evaluate the suitability of portable NIR sensors (food-scanners) for the determination of fruit quality along the supply chain of fruit and vegetables. This thesis focused on tomato as model fruit for the development of non-destructive prediction models of a variety of important quality attributes and evaluated the practicability of already existing and commercially available portable and miniaturized food-scanners.

Given this background information, the research questions for this doctoral thesis were formulated as follows:

- Miniaturized NIR instruments (food-scanners) are relatively new to the scientific literature. Is their performance comparable to traditional laboratory benchtop NIR instruments for the prediction of fruit quality traits?
- Currently there are various different food-scanners commercially available. Are there major differences with respect to the predictability of fruit quality parameters?

- With regard to the implementation of these novel devices in processes of quality control along the fresh fruit supply chain, are there differences with respect to the practicalness of commercially available food-scanners?
- Due to their working principle, food-scanners could determine multiple internal fruit quality parameters non-destructively. Is there a demand for these devices from the fresh produce industry?
- Can these food-scanners, which have primarily been used in this doctoral thesis to investigate the performance on predicting tomato fruit quality traits, also be used for a wide-ranging applicability of various types of fruit and vegetables and processes along the FSC?

Based on these research questions, the following hypotheses were phrased for this doctoral thesis:

H1. Actors along the FSC see a great potential for the application of food-scanners;
H2. There are different areas of application for food-scanners along the FSC;
H3. Prediction models derived from food-scanner spectra show comparable results than those computed with benchtop NIR instruments;
H4. Commercially available food-scanners achieve similar prediction model accuracy of selected fruit quality parameters;
H5. Software solutions provided by some food-scanner manufacturers achieve prediction model performance comparable to state-of-the-art multivariate analysis software
H6. Currently available devices differ in practicability and applicability with respect to the requirements of the fruit and vegetable supply chain;
H7. Food-scanners can be used for the prediction of various quality-related parameters on a wide range of fruit and vegetables;

1.2 Thesis structure and objectives

This thesis is structured into nine chapters, which comprise the objectives set for this study in order to answer the above stated research questions, and to corroborate the derived hypotheses (Figure 2):

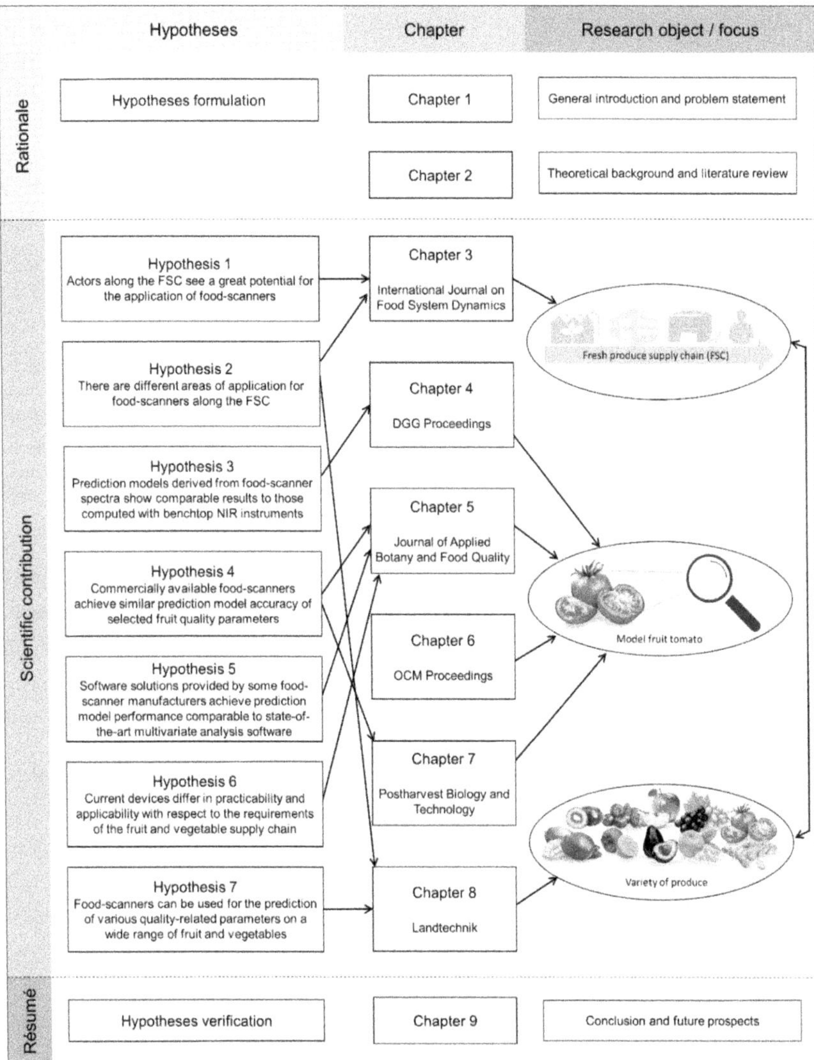

Figure 2: Overview of the structure of this doctoral thesis

Chapter 1 serves as introduction and states the status quo and motivation of this thesis. Additionally, the research questions as well as the hypotheses of the doctoral thesis are outlined.

Chapter 2 provides the theoretical background on near infrared spectroscopy and its application within the field of fruit and vegetables. Additionally, the development from laboratory to portable and miniaturized sensors is outlined and an extensive literature review of scientific research utilizing these new devices provided.

Chapter 3 investigates the status quo on how measurement of fruit quality is currently performed on various levels along the fruit and vegetable supply chain. Furhtermore it presents the opinion of experts along the supply chain on the application of food-scanners for non-destructive quality assessment and thereby justifies the basis for this work. This work has been published in 2020 in the International Journal on Food System Dynamics 11 (2), pages 101-106.

Chapter 4 compares the prediction accuracy of a miniaturized food-scanner and a conventional laboratory spectrometer using selected quality parameters of tomato fruit. The results of this work have been presented at the Horticultural Science Conference in Geisenheim 2018 and were issued in the DGG Proceedings 8 (13), pages 1-5.

Chapter 5 evaluates the capability and practical suitability of three commercial food-scanners for non-destructive measurement of important quality parameters using the model fruit tomato. Additionally, prediction models derived from software solutions provided by manufacturer are compared to state-of-the-art software for multivariate analysis. This chapter furthermore illustrates the possibility of utilizing food-scanners for quality control of pre- and postharvest processes. The findings have been published in 2020 in the Journal of Applied Botany and Food Quality 93, pages 204-214.

Chapter 6 examines the possibility of shelf-life assessment of tomatoes using a miniaturized NIR device. For this purpose, storage trials were carried out in combination with NIR measurements on different tomato varieties. Mathematical models were computed to estimate the tomato firmness over the storage period. The findings of this research have been presented at the 4th International Conference on Optical Characterization of Materials in Karlsruhe in 2019 and were subsequently published in the OCM Proceedings, pages 1-12.

Chapter 7 investigates the measurement of secondary plant metabolites with the help of food-scanners using lycopene content in tomato as example. Three commercially available food-scanners are tested and compared with respect to their accuracy and performance to each other as well as to alternative measurements of lycopin by using color values derived from a colorimeter. The results have been issued in 2020 in Postharvest Biology and Technology 167 (111232).

Chapter 8 evaluates the performance of three portable and miniaturized food-scanners with regard to their predictive accuracy of important quality parameters on a wide range of produce from the fruit and vegetable assortment. Additionally, results of a first in-field practical application of a food-scanner for quality control during incoming goods control are presented and further areas of application are discussed. The findings of this work have been published in 2021 in the issue (76) 1 of the journal Landtechnik, pages 52-67.

Chapter 9 provides a comprehensive discussion and links the hypothesis to the findings obtained in this thesis. Future prospects of food-scanner applications along the supply chain and remaining challenges for their implementation are exposed.

2 Theoretical background and literature review

2.1 Operating principle of near infrared spectroscopy

The near infrared (NIR) radiation is located between the visible and mid infrared part of the electromagnetic spectrum (Figure 3), which relates to the wavelength range of 780 to 2500 nm (Sandorfy et al. 2007). NIR radiation, due to its nature, acts like a wave and can be described with the two properties frequency and wavelength. NIR directed towards an object (e.g., liquid, solid) is absorbed by molecules, which as a result leads to the vibration of chemical bonds between atoms within these molecules (Osborne 2006). In a simplified approximation, these vibrations react as a harmonic motion, which can be described using a diatomic oscillator (Osborne 2006). Based on this simple vibrating system, additional concepts have been developed to more complex polyatomic molecules. The fundamentals and physical principles of these oscillating motions of di- and polyatomic molecules have been described in the literature in great detail (Bokobza 2002; Sandorfy et al. 2007).

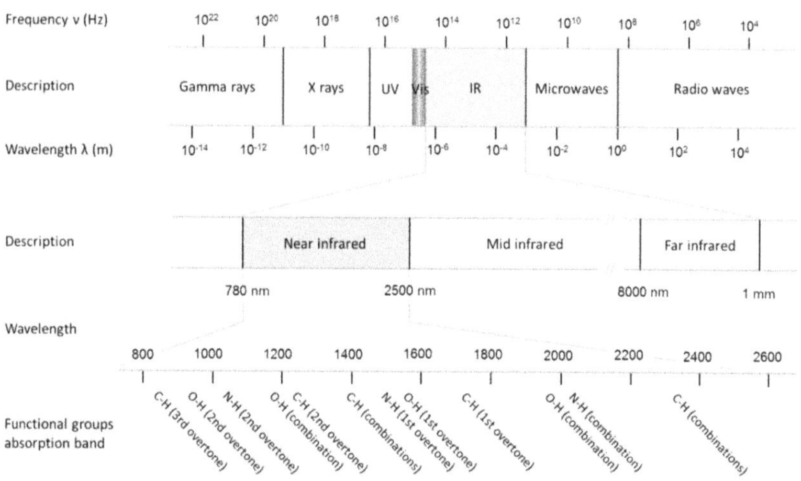

Figure 3: Location of near infrared radiation on the electromagnetic spectrum and approximate locations of absorption bands of important functional groups (own illustration based on Osborne (2006), Xiaobo et al. (2010))

Samples containing organic matter display distinct spectral fingerprints in the NIR region due to relatively prominent absorption of overtones and combination modes related to several functional groups (Figure 3) typically present in organic compounds, e.g., aliphatic and aromatic (C-H), carboxyl (C-O), hydroxyl (O-H) and amine and amide (N-H) groups (Xiaobo et al. 2010). Therefore, NIR is sometimes called the “overtone region” (Sandorfy et al. 2007). Due to the chemical composition of organic molecules and their particular absorption patterns in the NIR region, NIR radiation is suitable for the analysis of organic material and samples. Provided that the frequency of the radiation matches the frequency of the vibrating molecule, a transfer of energy from radiation to the molecule occurs, which subsequently can be measured and displayed as a plot of energy vs. wavelength, called a spectrum (Osborne 2006).

In some cases, in addition to NIR radiation, light visible to the human eye (VIS) is used for spectroscopic applications. In that case it is referred to as VIS/NIR spectroscopy. For reasons of simplification, only the term NIR spectroscopy will be used in the following sections.

2.2 Measurement modes in NIR spectroscopy

With regard to the sample composition and the trait of interest for analysis there are four different modes of measurement available in NIR spectroscopy: transmittance, reflectance, interactance and transflectance (Figure 4). The main operation principles and differences between these modes are as follows (Tsuchikawa 2007; Pasquini 2003): In transmission mode, the sample (S) is illuminated perpendicular from one side by a light source (L) and the transmitted light is subsequently detected on the opposite side of the sample with a detector (D). In a similar manner, the sample is illuminated perpendicular to the sample surface in reflection mode. However, in contrast to transmission, the reflection mode employs reflected light from the sample surface, in which the incident light, after illumination, underwent various effects such as scattering, absorption and transmission. Reflection measurements are oftentimes conducted in a specific distance (d) between illumination source and sample, whereas the reflected light is in general collected by the detector in a fixed angle (e.g., 45°) and distance to the sample.

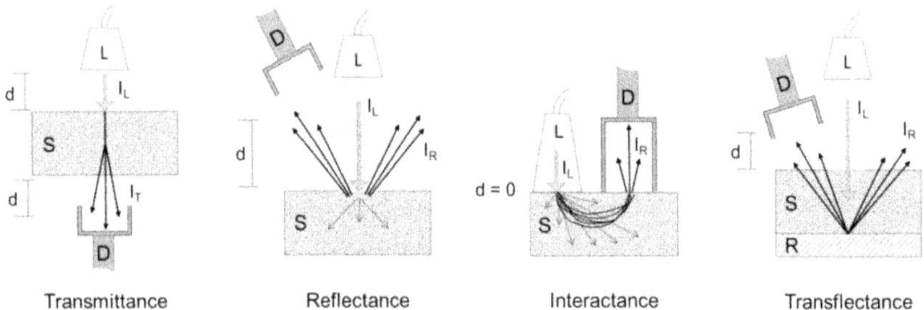

Figure 4: Different modes of measurement used in NIR spectroscopy, where I_L = incident light, I_T = transmitted light, I_R = reflected light, D = detector, R = reflector, S = sample, d = distance (own illustration based on Tsuchikawa (2007), Pasquini (2003), and Saranwong and Kawano (2007))

Interactance mode usually makes use of a specifically designed interaction probe, which generally consists of an illuminator and detector both in direct contact with the sample surface but arranged in certain distance to each other. Usually the light source and detector are positioned in a parallel setup so that the light, due to specular reflection, cannot directly be detected (Nicolaï et al. 2007). As a result, only the light directly transmitted through the sample is detected, and more information about the actual sample composition and sample constituents can be obtained (Pasquini 2003). The transflection mode constitutes a combination of transmission and reflection, where the illuminated light, after passing through the sample, is scattered back from a reflector (R) and passed through the sample a second time (Tsuchikawa 2007; Pasquini 2003). The reflected light is then collected by a detector closely positioned to the sample. A detailed explanation of the use of the different NIR measurement methods for agricultural and horticultural products analysis can be found in the following chapter 2.5.

2.3 Chemometrics

Due to the complexity of organic material, the crucial information present in NIR spectra about the chemical composition is scattered over the whole spectrum and generally not directly available for analytical purpose. The evaluation of NIR spectra requires a special type of data analysis, which simultaneously recognizes changes in the whole or part of the spectrum as a function of the changes in the sample properties or/and analytes contents (Pasquini 2018). This technique is called multivariate analysis or chemometrics (Brown 2016; Cozzolino 2009), which since its development in the middle of the 1970 became "one of the sustaining pillars of modern NIRS" (Pasquini 2018). The application of chemometric techniques helps to obtain quantitative as well as qualitative relations between NIR spectra and the trait of interest (Figure 1).

Partial least squares regression (PLSR), a method which assumes a linear relationship between the spectral vibration and the modeled parameter or concentration (Pasquini 2018), is commonly used for quantitative evaluations, e.g., the estimation of soil element contents (Yu et al. 2016), the determination of the concentration of quality parameters in alcohol (Valderrama et al. 2007) and the analysis of sugar, acid and phenol content of fruit (Guo et al. 2016). In recent years, several additional techniques based on PLSR algorithm have been developed, e.g. nonlinear iterative partial least squares (NIPALS), sparse partial least squares (SPLS) and mixed-norm partial least squares (MNPLS) (You et al. 2016).

Qualitative evaluations are used in order to classify and authenticate product origin and quality as well as identify product adulterations and falsifications. The most prominent methods include linear discriminant analysis (LDA), cluster analysis (CA), soft independent modeling of class analogy (SIMCA), k-nearest neighbor (KNN), support vector machine discriminant analysis (SVM-DA), partial least squares discriminant analysis (PLS-DA), and principal component analysis-artificial neural networks (PCA-ANN) as well as improved and derived versions thereof (Pasquini 2018). Within the last decade, these methods have been applied for various purposes in different research fields related to NIR spectroscopy, e.g., the forensic classification of pen ink and printer inkjet (Silva et al. 2013; Oravec et al. 2019), the detection of adulterants in honey (Zhu et al. 2010) and milk powder (Mazivila et al. 2020), the classification of cereal bars (Brito et al. 2013), and the identification of herbal medicines (Yang et al. 2013).

The focus of this work was on quantitative analyses for fruit quality parameters. For this purpose, PLSR alogorithm was used for the correlation of the spectra derived from food-scanners to fruit quality parameters by using the independent multivariat analysis software The Unscrambler®. The software provided by the manufacturer of the SCiO™ device uses PLSR algorithm as default method. According to the user manual provided by Felix Instruments the provided software Model Builder of the F-750 uses NIPALS for the correlation of spectra and reference values (Felix Instruments 2020a). Since the H-100F device is not provided with an individual software, spectra was analyzed using The Unscrambler®.

2.4 Development of NIR prediction models

In general, the development of feasible NIR prediction models follows a specific procedure (Figure 5), which has been elaborately described in the relevant literature (Cozzolino 2020; Pasquini 2003; Næs et al. 2004).

In order to produce a robust prediction model, the selected samples should represent the natural variability of the property or concentration of interest (Pasquini 2003). In case of fruit quality, the robustness of models could therefore be improved by combining several cultivars of fruit of the same species which are for example characterized by different degrees of sweetness, or incorporating fruit of different origin within one model. The robustness of the prediction model as well as the accuracy of prediction is tested during the final step of model building, model validation. A good validation is characterized by the fact that it uses external samples that have not been used for cross-validation and that have been sampled and analyzed separately (Cozzolino 2020). For fruit this could entail using samples of the same species but of different cultivars, origins or cultivation methods (e.g. organic, conventional).

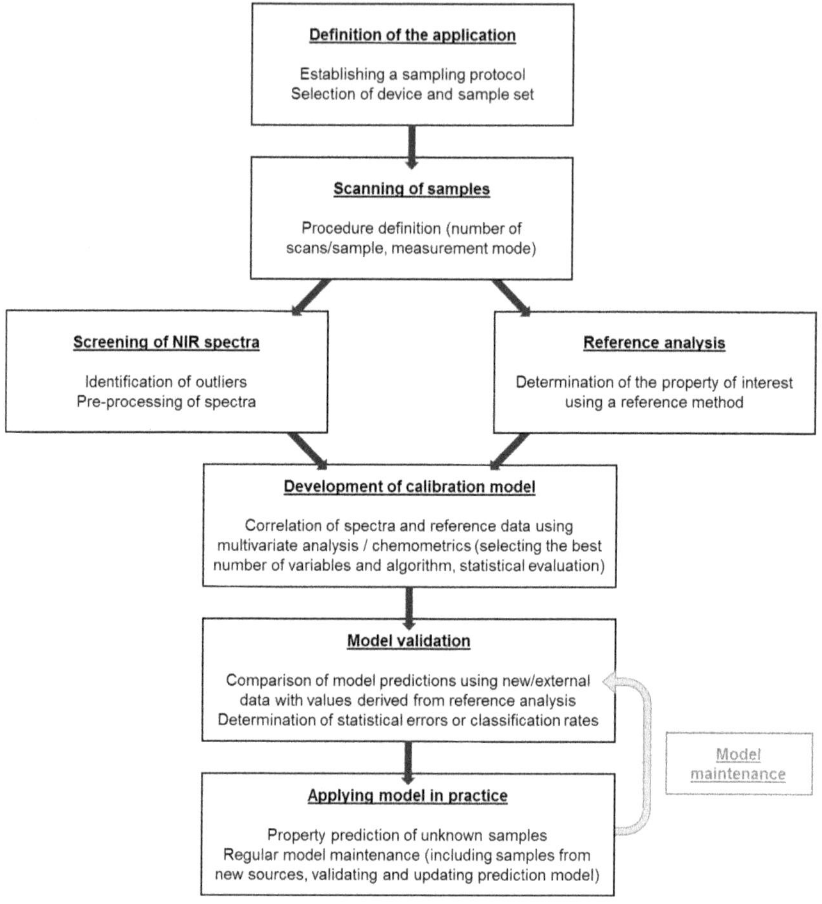

Figure 5: General procedure in planning, developping, evaluating, utilizing and maintaining of NIR prediction models (own illustration based on Cozzolino (2020), Pasquini (2003)).

2.5 NIR applications for agricultural and horticultural products

Although radiation in the NIR region was discovered as early as 1800 (Herschel 1800), it was not until Karl Norris' work in the 1960s that this spectroscopic method was first used to analyze chemically complex samples (Nicolaï et al. 2007; McClure 1994). Norris, who is referred to by many authors as the "modern mentor" (McClure 1994) and "pioneer researcher" (Pasquini 2003) of NIR spectroscopy, examined various agricultural applications in his pioneering work, e.g., the spectrophotometric determination of moisture content of grain and seeds (Norris and Hart 1965), the prediction of forage quality (Norris et al. 1976), and the analysis of the chemical composition of tobacco (McClure et al. 1977). Due to his adoption of a multiple-wavelength model for the correlation of spectral data to reference values he was the first researcher to recognize the potential for enabling quantitative measurements from NIR spectra using multivariate analysis (McClure 1994; Pasquini 2003).

As summarized by Nicolaï et al. (2007), the first applications of NIR spectroscopy with respect to horticultural products focused on the prediction of internal fruit and vegetable quality parameters such as the dry matter content of onions (Birth et al. 1985), the soluble sugar content (SSC) of intact peaches (Kawano et al. 1992) and mandarins (Kawano et al. 1993) and the moisture content of mushrooms (Roy et al. 1993). Starting from these experiments, there has been a growing interest in utilizing NIR spectroscopy for the detection of fruit quality, especially within the last 15 years. The growing number of scientific publications per year related to this topic can be illustrated through a keyword-based search on Google Scholar (Figure 6). The abbreviation "NIR" was used as a fixed keyword in all searches and combined with terms of relevance for this thesis, e.g., "fruit quality" and specific internal fruit quality parameters such as "sugar content" or "dry matter". The results in numbers of this search can be found in the appendix (Chapter 10).

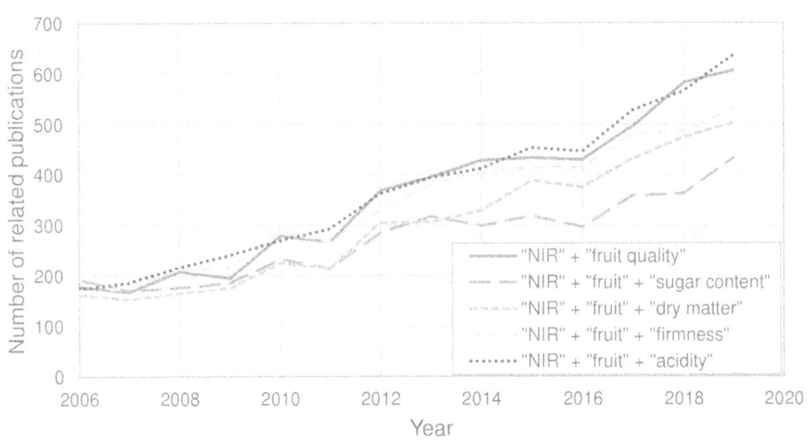

Figure 6: Number of publications per year related to keywords including "NIR" and terms associated to fruit quality and selected fruit quality parameters found on Google Scholar in December 2020

As indicated by the literature, NIR analysis within the field of fruit and vegetables has been used to a great extent to identify quantitative chemical traits such as sugar content, dry matter/moisture content and acidity of fruit. Additionally, NIR has been employed for the detection of physical traits, e.g., firmness of produce.

In addition to these generally important and relatively easily identifiable quality parameters in fruit and vegetable, the scope for the utilization of NIR spectroscopy has been further extended to the detection of secondary plant metabolites in these products. For fruit, applications include the measurement of vitamin C and polyphenol content in apples and kiwifruit (Beghi et al. 2013; Fu et al. 2008; Pissard et al. 2013), the determination of total flavonoids and anthocyanins in blueberries (Bai et al. 2014; Sinelli et al. 2008), assessment of phenolic content in strawberries (Włodarska et al. 2019) and grapes (Ferrer-Gallego et al. 2011), the prediction of oil content in avocado (Ncama et al. 2018; Olarewaju et al. 2016) and the quantification of lycopene and β-carotene in watermelons (Tamburini et al. 2017). For vegetables, similar applications for the measurement of secondary plant metabolites have been investigated, e.g., the detection of chlorophyll and phenolic compounds in summer squash (Blanco-Díaz et al. 2014), the determination of carotenoids in carrots (Schulz et al. 1998) as well as lycopene in tomatoes (Clément et al. 2008; Li et al. 2017b), the prediction of the total phenolic content and antioxidant capacity of onion and shallot (Lu et al. 2011) and the screening of individual and total glucosinolate contents in broccoli (Hernández-Hierro et al. 2012). Additional studies found great potential in NIR spectroscopy for non-destructively tracing fruit cultivar and orchard origin (Eisenstecken et al. 2019) and the separation of quality classes with respect to pest-infection (Saranwong et al. 2010), bruises (Xing et al. 2005; Zhang et al. 2013) fruit maturity (Khodabakhshian et al. 2017; Sirisomboon et al. 2012) and sensory attributes (Mehinagic et al. 2003; Parpinello et al. 2013) of intact fruit. A summary of potential NIR applications for quantitative and qualitative analysis with respect to fresh produce is provided in Table 1.

Table 1: Potential VIS/NIR applications for fruit and vegetables

Type of analysis	Quantitative	Qualitative
Examples	Determination of chemical properties • Sugar content • Dry matter / moisture • Acidity / pH • Secondary plant metabolites Assessmment of physical properties • Firmness • Color value	Sorting / classification of fruit by • Cultivar • Origin • External defects / bruises • Internal defects / browning • Pest-infection • Pre-definded maturity classes • Human-sensory attributes • Color grading

With regard to the application of NIR spectroscopy and the measurement of fruit, transmittance, reflectance and interactance mode have been reported to obtain usable near infrared spectra (Nicolaï et al. 2007). In this context, transflectance mode only takes a minor role and can be disregarded, since it was originally designed to measure thin and/or clear samples such as liquids or turbid materials (Tsuchikawa 2007). As summarized by Saranwong and Kawano (2007), in the context of the application of these different modes for the measurement of fruit and vegetables, the thickness of the fruit peel can be used as indicator of which mode should be applied. Transmittance mode can help to gather information about the constituents of thick-peeled fruit (e.g., citrus fruit, cantaloupes, melons), however these measurements require very high light intensities, which could result in burning of fruit surface and the alteration of spectral information. Additionally, transmittance spectra can carry information about fruit skin and fruit core, which might be irrelevant to the actual measurement (Nicolaï et al. 2007). Nevertheless, a review on on- and in-line monitoring of food quality using NIR spectroscopy by Huang et al. (2008) refers to research using transmittance spectroscopy in apples (Clark et al. 2003) which illustrated the possibility of using transmittance spectra for the prediction of area affected by internal browning. Additional research indicates the possibility of using NIR transmittance spectra for the detection of brown heart of pear (Fu et al. 2007) and internal browning in pineapples (Sukwanit and Teerachaichayut 2013). However, it should be noted that conventional NIR measurements are measurements focused on a single point, which have their limits when it comes to detecting punctual damage within the fruit. Imaging techniques such as hyperspectral imaging could be the better choice for the detection of internal fruit damage, as they have a better chance of detecting such punctual damage due to their operating principle (e.g. using transmittance spectra, analyzing a two-dimensional image with a large number of pixels containing NIR spectral information) (Ariana and Lu 2010). Reflectance mode is mostly suitable for the detection of information located in or directly under the peel, e.g., anthocyanins in blueberries (Saranwong and Kawano 2007), since the light penetration in this mode of measurement is reported to be limited to a few millimeters, depending on the respective wavelength range (Lammertyn et al. 2001; Fraser et al. 2001; Lammertyn et al. 2000). The inhomogeneous composition and rough surface of fruit and vegetables leads to a change of the signal intensity due to scattering and absorbance of light, therefore reflectance measurements in this context are oftentimes referred to as diffuse reflectance (Pasquini, 2003). A more suitable mode for the measurement of thin-peeled fruit, and the overall “most popular mode for use on fruits and vegetables” (Saranwong and Kawano 2007) is the interactance mode. This mode oftentimes uses a specifically designed interactance probe and fruit holder, which enables a parallel positioning of light source and detector (Kawano et al. 1992) and allows the detection of information from deep within the fruit while excluding information reflected from the fruit surface (Saranwong and Kawano 2007). However, other designs than a parallel setup for the measurement in interactance mode have been used in scientific setups (Nicolaï et al. 2007), e.g., for the measurement of internal fruit quality parameters of apples (McGlone et al. 2002) and rockmelons (Greensill and Walsh 2000).

The three devices used in this thesis operate in interactance mode. Whereas the SCiO™ and

F-750 instrument apply a classical interactance measurement setup where the light source and detector are aligned parallel to each other, the interactance measurement of the H-100F device uses an angular placement of light source and detector.

2.6 Transition from stationary laboratory instruments to portable and miniaturized NIR sensors

Due to the above mentioned different types of applications of NIR spectroscopy in the field of fruit and vegetables (Table 1), various possibilities for the utilization of this technology emerged along the whole FSC. As highlighted by Nicolaï et al. in the year 2007, most applications by that time had been carried out under static conditions and in laboratory environments, and most commercial applications until then involved postharvest grading lines, with first concepts of multi- and hyperspectral online systems emerging. Within the literature, different monitoring and measurement strategies of postharvest produce are described with regard to the respective implementation process (Cortés et al. 2019; Dickens 2010), however scanning of fruit samples is generally performed in a laboratory setup utilizing a stationary NIR spectrometer (Table 2). In this context, food-scanners can be used for "off-line" and "at-line" measurement of fruit quality.

Table 2: Terms and definitions of different postharvest monitoring strategies

Type of measurement	Definition
"on-line"	Samples are rerouted from the original production line and analyzed directly in a by-pass line, after analysis samples are recirculated to the production line
"off-line"	Samples are analyzed remote from the production line, e.g. within a laboratory
"at-line"	Samples are manually and randomely removed from the production line, analysis is performed closely nearby
"in-line" or "*in situ*"	Samples are analyzed within the running production line

(Source: Cortés et al. (2019); Dickens (2010))

Due to its non-destructive nature, the VIS/NIR technology can be used to determine quality before harvest. Portable VIS/NIR devices, which have been developed for this purpose in recent years, are particularly suitable for this application. Promising results, among others, have been achieved for the determination of on-tree fruit quality of apples (Geyer et al. 2007), avocados (Ncama et al. 2018), apricots (Camps and Christen 2009), peaches and nectarines (Costa et al. 2002; Pérez-Marín et al. 2009), mangos (Saranwong et al. 2003) and tomatoes (Kusumiyati et al. 2008) using portable VIS/NIR instruments. Some of these studies include elaborate setups in order to enable the operation of the spectrometer, e.g., the transport of a

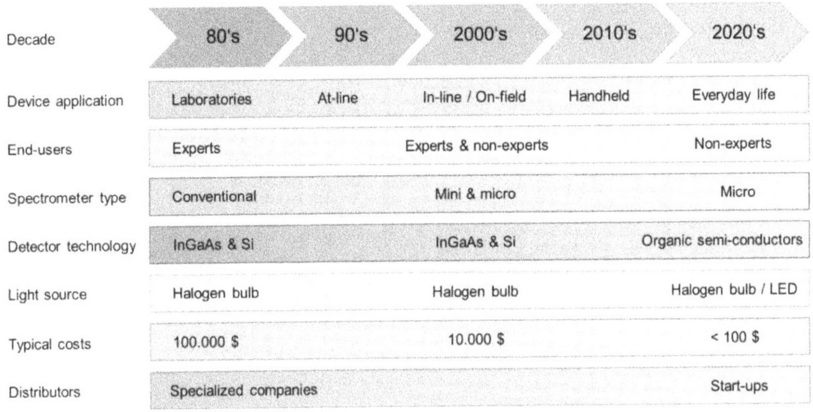

Decade	80's	90's	2000's	2010's	2020's
Device application	Laboratories	At-line	In-line / On-field	Handheld	Everyday life
End-users	Experts		Experts & non-experts		Non-experts
Spectrometer type	Conventional		Mini & micro		Micro
Detector technology	InGaAs & Si		InGaAs & Si		Organic semi-conductors
Light source	Halogen bulb		Halogen bulb		Halogen bulb / LED
Typical costs	100.000 $		10.000 $		< 100 $
Distributors	Specialized companies				Start-ups

Figure 7: The evolution of spectrometers from laboratory instruments towards end-consumer devices (own illustration based on Beć et al. (2020), McClure (1994), Senorics (2020), and Spectral Engines Oy (2018))

claim to be complete. Furthermore, this summary is limited to handheld, portable and miniaturized devices that function as a unit and may be operated in combination with a smartphone or tablet. Studies that use portable spectrometers in an elaborate laboratory and de facto not portable setup (e.g., the USB2000+ spectrometer distributed by Ocean Optics as utilized by Fan et al. (2020)) were omitted from this summary.

Table 4: Overview of scientific studies using portable, handheld and miniaturized VIS/NIR spectrometers for quality detection in fruit and vegetables (2015-2021)

Fruit species	Studied parameter(s)	VIS/NIR instrument(s)	Reference
Apple	Firmness, total soluble solids, titratable acidity, starch-iodine index	MicroNIR	Pissard et al. (2021)
	Dry matter	F-750	Teh et al. (2020)
	Dry matter, total soluble solids	F-750	Zhang et al. (2020)
	Classification of postharvest treatment	MicroNIR	Buccheri et al. (2019)
	Dry matter, total soluble solids	F-750	Zhang et al. (2019)
	Dry matter, firmness, total soluble solids	SCiO™	Li et al. (2018)
	Dry matter	Nirvana, F-750, H-100C, SCiO™	Kaur et al. (2017)
	Total soluble solids	K-BA100R	Qi et al. (2017)
	Chlorophyll content	Prototype	Das et al. (2016)
	Classification of sound and defect fruit	Nirvana	Khatiwada et al. (2016)
	Total soluble solids, polyphenol content	MicroPhazir	Schmutzler and Huck (2016)
	Dry matter, total soluble solids	Nirvana	Kumar et al. (2015)
	Total soluble solids	K-BA100R	Liu et al. (2015)
Acerola	Titratable acidity, ascorbic acid	MicroNIR	Malegori et al. (2017)
Avocado	Dry matter	F-750, MicroNIR, SCiO™	Subedi and Walsh (2020)
	Firmness, ripeness stage	SCiO™	Li et al. (2018)
	Dry matter, oil content, moisture content	F-750	Ncama et al. (2018)
	Moisture content	Phazir 1018	Blakey (2016)
Bananito fruit	Total soluble solids	MicroNIR	Pu et al. (2018)
Bell pepper	Dry matter, total soluble solids, titratable acidity, color index	MicroNIR	Sánchez et al. (2020)
	Dry matter, total soluble solids, authentication by growing method	MicroPhazir	Sánchez et al. (2019)
	Firmness, total soluble solids, pH, malic acid, pericarp wall thickness, skin color	Phazir 1018	Toledo-Martín et al. (2016)
Bitter melon	Firmness, total soluble solids, skin color	Nirvana	Kusumiyati et al. (2018b)

Table 4 *(Continued)*

Fruit species	Studied parameter(s)	VIS/NIR instrument(s)	Reference
Cashew Apple	Total soluble solids, titratable acidity, ascorbic acid, sugar/acid ratio	MicroNIR	Ribeiro et al. (2016)
Citrus fruit	Total soluble solids, titratable acidity, vitamin C, classification of fruit type	Tellspec®	Santos et al. (2020)
	Total soluble solids, pH, titratable acidity, maturity index, BrimA index	Phazir 2400	Torres et al. (2019b)
	Color values, firmness, pericarp thickness, juice mass	Phazir 2400	Torres et al. (2017)
Cucumber	Water content, total soluble solids, firmness	Nirvana	Kusumiyati et al. (2020)
Feijoa	Skin color, firmness, total soluble solids, titratable acidity	SCiO™	Li et al. (2018)
Gandaria	Total soluble solids	MicroNIR	Posom et al. (2020)
Garlic	Dry matter, total soluble solids	Nirvana	Jantra et al. (2017)
Grape	Macro-element content	MicroNIR	Cuq et al. (2020)
	Dry matter, total soluble solids	F-750, SCiO™	Donis-González et al. (2020)
	Phenolic content, flavanol content, anthocyanin content	MicroNIR	Baca-Bocanegra et al. (2019)
	Dry matter, total soluble solids	SCiO™	Momin et al. (2019)
Kiwifruit	Total soluble solids	F-750	Sarkar et al. (2020)
	Dry matter, total soluble solids	Prototype	Singh et al. (2020)
	Dry matter, firmness, total soluble solids	SCiO™	Li et al. (2018)
	Dry matter	Nirvana, F-750, H-100C, SCiO™	Kaur et al. (2017)
Mango	Total soluble solids	DLP® NIRscan™ Nano	Amirul et al. (2020)
	Dry matter	F-750	Anderson et al. (2020)
	Internal browning	F-750	Gabriëls et al. (2020)
	Firmness	F-750	Mishra et al. (2020)
	Internal disorders	F-750	Mogollón et al. (2020)
	Dry matter	F-750	Sun et al. (2020b)

Table 4 *(Continued)*

Fruit species	Studied parameter(s)	VIS/NIR instrument(s)	Reference
Mango	Moisture content	F-750	Wokadala et al. (2020)
	Total soluble solids	DLP® NIRscan™ Nano	Al-Sanabani et al. (2019)
	Dry matter	F-750	dos Santos Neto et al. (2018)
	Dry matter	F-750	Anderson et al. (2017)
	Dry matter, total soluble solids	F-750	dos Santos Neto et al. (2017)
	Detection of artificial ripening	MS-720	Mithun et al. (2017)
	Dry matter, total soluble solids, titratable acidity, pulp firmness	MicroNIR	Marques et al. (2016)
Melon	Total soluble solids	K-BA100R	Hu et al. (2019)
	Firmness, total soluble solids	K-BA100R	Lu et al. (2015)
Mulberry fruit	Dry matter, total soluble solids, polyphenol content, flavonoid content	MicroNIR	Yan et al. (2019)
Olive	Dry matter	F-750	Sun et al. (2020a)
Onion	Dry matter, total soluble solids	Nirvana	Jantra et al. (2017)
Oranges	Weight, firmness, total soluble solids, titratable acidity, BrimA index	Phazir 2400	Pérez-Marín et al. (2019a)
Peach	Dry matter, total soluble solids, flesh firmness, index of absorbance difference	F-750	Minas et al. (2021)
	Dry matter, total soluble solids	F-750, SCiO™	Donis-González et al. (2020)
	Flesh firmness, water soluble pectines	K-BA100R	Uwadaira et al. (2018)
	Total soluble solids	K-BA100R	Liu et al. (2015)
Pear	Moisture content, total soluble solids	F-750, DLP® NIRscan™ Nano	Mishra et al. (2021a)
	Moisture content, total soluble solids	F-750	Mishra et al. (2021b)
	Dry matter	F-750	Goke et al. (2020)
	Total soluble solids	DLP® NIRscan™ Nano	Yuan et al. (2020)
	Dry matter	F-750	Serra et al. (2019)

Table 4 *(Continued)*

Fruit species	Studied parameter(s)	VIS/NIR instrument(s)	Reference
Pear	Dry matter, total soluble solids	F-750	Goke et al. (2018)
	Total soluble solids	H-100F	Choi et al. (2017)
	Total soluble solids, firmness	K-BA100R	Wang et al. (2017)
	Total soluble solids	K-BA100R	Liu et al. (2015)
Persimmon	Firmness	DLP® 2010 NIR	Su et al. (2020)
Pineapple	Total soluble solids, classification organic / conventional	SCiO™	Amuah et al. (2019)
Plum	Classification of cultivar, growing location and anthocyanin content	MicroNIR	McIntyre et al. (2020)
	Firmness, total soluble solids, pH, titratable acidity, sugar /acid ratio, skin color	K-BA100R	Li et al. (2017a)
Ridge Gourd	Moisture content, firmness, total soluble solids, skin color	Nirvana	Kusumiyati et al. (2017)
Sapodilla	Moisture content, firmness, total soluble solids, skin color	Nirvana	Kusumiyati et al. (2019)
	Firmness, total soluble solids, skin color	Nirvana	Kusumiyati et al. (2018a)
Spinach	Dry matter, total soluble solids, nitrate content	MicroNIR	Torres et al. (2021)
	Dry matter, total soluble solids, nitrate content, maximum puncture force	MicroNIR	Entrenas et al. (2020)
	Total soluble solids, nitrate content	MicroNIR	Torres et al. (2020b)
	Total soluble solids, ascorbic acid, nitrate content	Phazir 2400	Pérez-Marín et al. (2019b)
Stone fruit	Dry matter, total soluble solids, flesh firmness	F-750	Scalisi and O'Connell (2020)
	Dry matter	Nirvana, F-750, H-100C, SCiO™	Kaur et al. (2017)
Summer squash	Dry matter, total soluble solids, nitrate content	MicroNIR	Torres et al. (2020a)
	Dry matter, total soluble solids, nitrate content	Phazir 2400, MicroPhazir, MicroNIR	Entrenas et al. (2019)
	Dry matter, total soluble solids, nitrate content	MicroNIR	Torres et al. (2019a)
	Weight, length, equatorial diameter, skin color	Phazir 2400	Sánchez et al. (2018)
	Dry matter, total soluble solids, firmness, pH, titratable acidity, nitrate content	Phazir 2400	Sánchez et al. (2017)

Table 4 *(Continued)*

Fruit species	Studied parameter(s)	VIS/NIR instrument(s)	Reference
Sweet cherry	Total soluble solids	DLP® NIRscan™ Nano	Wang et al. (2018)
	Dry matter, total soluble solids	F-750	Escribano et al. (2017)
	Dry matter	F-750	Toivonen et al. (2017)
Tomato	Dry matter, total soluble solids, pH, titratable acidity, skin color, electrical conductivity	MicroNIR	Castrignanò et al. (2019)
	Total soluble solids, lycopene content	DLP® NIRscan™ Nano	Sheng et al. (2019)
	Dry matter, color values	Nirvana	Acharya et al. (2017)
	Total soluble solids, lycopene content	DLP® 2010 NIR	Zhiming et al. (2017)
	Classification by storage time	SCiO™	Lee et al. (2017)
Umbu	Dry matter, total soluble solids, flesh firmness, skin color	F-750, SCiO™, Tellspec®	Marques and Freitas (2020)
Watermelon	Classification by storage time	SCiO™	Lee et al. (2017)

2.7 Exploring the potential of food-scanners using tomato as model fruit

At the beginning of this thesis in 2017, the literature on miniaturized scanners and fresh produce was not yet as numerous (Table 4). In order to investigate the potential of food-scanners to determine fruit quality along the FSC, a fruit species that has not yet been extensively researched was selected as model fruit. As this work focused primarily on the German FSC, fresh tomato (*Solanum lycopersicum* L.) was selected as model fruit due to its high importance in the German market and the large area under cultivation. With a quantity of 2.3 Mio t per year, tomato is the most widely consumed vegetable in Germany (Statista 2020). Additionally, with 30% of Germany's greenhouse production area, which equals a total of 398 ha, tomato occupies the first place of vegetables grown under glass (Illert 2019).

The tomato fruit is also very well suited as a test object for food-scanners due to its numerous quality attributes. During on-vine ripening the metabolic processes cause changes in the texture and quality of tomato, e.g., a decrease in firmness, increase in lycopene and shift of color, accumulation of sugars and changes in the content of organic acids (Brandt et al. 2006; Bertin and Génard 2018; Oms-Oliu et al. 2011). Furthermore, tomato is climacteric (Chalmers and Rowan 1971), and the increase of ethylene production rates at the time of ripening (Ku et al. 1970; Oms-Oliu et al. 2011) results in further changes of tomato quality during postharvest storage (Javanmardi and Kubota 2006; Toor and Savage 2006; Majidi et al. 2014). Results derived from experiments conducted at the University of Applied Sciences Weihenstephan-Triesdorf illustrate the changes in tomato fruit quality pre- and postharvest (Figure 8).

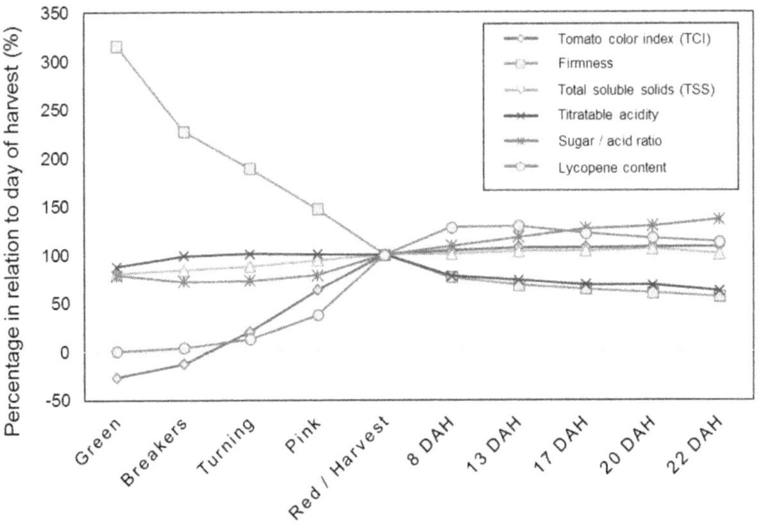

Figure 8: Changes in tomato quality parameters during fruit ripening and postharvest storage at 22° C

With respect to shelf life and consumer acceptance, fruit color and firmness have been identified as important quality parameters (Batu 2004). Furthermore, sensory traits like sweetness, taste intensity, and juiciness, which vary between tomato cultivars due to differences in sugar and acidity content, have been reported to influence consumers' purchase preference (Casals et al. 2019).

As indicated by these researches, tomato quality is a multi-faceted trait and therefore can not be described using only selected parameters. Normally, the determination of the most important fruit parameters requires a variety of different methods and instruments, such as the measurement of fruit firmness via penetrometer, the determination of sugar content using a refractometer or the analysis of titratable acidity by titration. Usually during quality control, several fruits are randomly and destructively analysed and the values determined are used as an indication of the quality of the whole batch (OECD 2018). As emphazised in the "Guidelines on objective tests" for the quality determination of fresh produce, the spectral signature of fruit and vegetables contains several dominant absorption bands in the VIS/NIR region (OECD 2018). Therefore, constituents such as chlorophyll (670 nm), red-colored pigments (500-600 nm), water (970 nm) as well as sugar and other carbohydrates can be detected in the VIS/NIR region employing spectroscopic measurements. In case of tomato, the potential of using this technology for the distinction of fruit of different quality and ripening stages is illustrated in Figure 9.

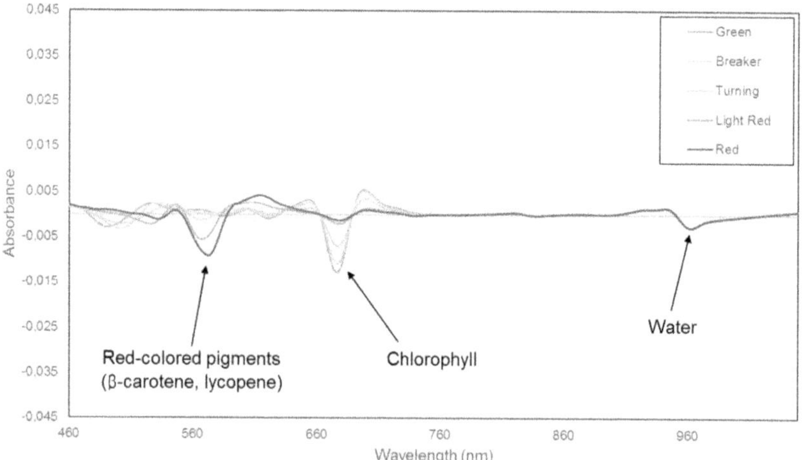

Figure 9: Averaged VIS/NIR tomato fruit spectra of different ripening stages including prominent absorption bands of red-colored pigments, chlorophyll, and water (own illustration based on F-750 measurements)

Studies regarding the predictability of tomato quality using laboratory VIS/NIR instruments yielded good results for various quality traits such as firmness, dry matter, sugar content, pH, skin color and lycopene content (Clément et al. 2008; Radzevičius et al. 2016; Shao et al. 2007). In addition, studies published at the beginning of this thesis indicated the potential of utilizing portable and miniaturized VIS/NIR instruments for these measurements (Table 4).

Based on this knowledge, the performance and suitability of portable and miniaturized devices for predicting fruit quality along the FSC has been investigated. Initially, tomato was used as model fruit and multiple important quality traits were examined. Subsequently, the established measurement procedures were transferred to other types of produce, and applications of food-scanners as non-destructive measurement tools along the FSC, such as the utilization in processes of quality control during incoming goods control, examined.

3 Food-scanners as a radical innovation in German Fresh Produce Supply Chains

In: International Journal on Food System Dynamics 11 (2), p. 101-116

Cite as: Goisser, S.; Mempel, H.; Bitsch, V. (2020): Food-Scanners as a Radical Innovation in German Fresh Produce Supply Chains. International Journal on Food System Dynamics 11 (2), p. 101-116.

DOI: https://doi.org/10.18461/ijfsd.v11i2.43

Title

Food-scanners as a radical innovation in German fresh produce supply chains

Author names and affiliations

Simon Goisser[a], Heike Mempel[a], Vera Bitsch[b]

[a]University of Applied Sciences Weihenstephan-Triesdorf, Greenhouse Technology and Quality Management, Am Staudengarten 10, 85354 Freising, Germany
[b]Technical University of Munich, School of Management & School and Life Sciences Weihenstephan, Chair of Economics of Horticulture and Landscaping, Alte Akademie 16, 85354 Freising, Germany

simon.goisser@hswt.de, heike.mempel@hswt.de, bitsch@tum.de

Corresponding author

Simon Goisser, Am Staudengarten 10, 85354 Freising, simon.goisser@hswt.de

Abstract

Originally advertised as tools for end-consumers, portable food-scanners have recently reached a high level of awareness and show potential as instruments for quality assessment along fruit and vegetable supply chains. The current study explores preferences and concerns of chain actors regarding the implementation of this technology through semi-structured interviews. Results indicate that food-scanners could facilitate quality control at different levels of the fresh produce supply chain by providing fast, non-destructive and objective measurements. Concerns about the application of food-scanners could be identified with respect to potential additional requirements of fruit wholesaler resulting in more pressure on producers. To further a goal-oriented and user-directed development of this new technology, future research should be directed at its impacts on perception of fruit quality along the chain as well as end-consumers' readiness to use these devices in everyday life.

Keywords

food-scanner; fruit and vegetable supply chain; quality measurement; qualitative research

1 Introduction

Quality and shelf life of produce depend on various product-specific parameters. Sugar content, brix/acid ratio, firmness and dry matter content are among the key parameters to determine maturity and ripeness of produce. Depending on the product, standards concerning the marketing and commercial quality control for some parameters must be met to allow distribution via retail chains (UNECE, 2017). In Germany, compliance with these standards are required from leading fruit retail companies. The verification of internal quality standards such as dry matter, sugar content and fruit acidity is often time-consuming and, in some cases, requires destructive measurement methods in combination with sample preparation and handling of chemicals (OECD, 2018). To optimize quality throughout the fresh food supply chain, degradation models and algorithms were developed and applied in several case studies for different fruit like bell peppers (Rong, Akkerman, & Grunow, 2011) and cantaloupes (Yu & Nagurney, 2013).

In recent years, so-called food-scanners became more and more popular for potential end-consumer applications, but also were tested in scientific studies. Food-scanners are mobile and miniaturized devices operating on the principle of near infrared (NIR) spectroscopy. Whereas traditional NIR spectrometers are expensive laboratory benchtop devices, portable devices apply the same operating principle for a fraction of the cost. Dos Santos, Lopo, Páscoa, and Lopes (2013) summarized commonly reported portable NIR instruments and applications for fruit and vegetable analysis in the literature and illustrate the main advantages of these devices as well as the possibility of using them under production conditions.

The potential for practical applications of portable food-scanners perceived by actors along the fresh produce supply chain (FSC) has not yet been studied. Since food-scanners constitute an innovation to the FSC, this study focuses on the new technology within the framework of innovation as well as drivers and barriers of innovation adoption. The cooperation between key stakeholders, in this case actors along the supply chain including researchers and companies developing food-scanners, is of high importance when it comes to adoption and impact of new technologies (Douthwaite, Keatinge, & Park, 2001). The present study therefore investigates the perspective of supply chain actors in Germany by highlighting the status quo in quality control and exploring potential NIR applications along the FSC. The objectives of the current study are to identify advantages and drawbacks as well as limitations of food-scanners as tools to complement traditional quality measurement methods. Since the adoption rate of new technologies depends on the motivation of key stakeholders to get to know the technology and to tailor it to their needs (Douthwaite et al., 2001) an additional objective is to identify the motivation of supply chain actors in applying and adjusting the food-scanner technology to their companies' requirements.

The structure of the paper is as follows: The next section presents a brief background on the portable food-scanner technology, followed by a literature review on innovation terminology as well as drivers and barriers for innovation adoption in chapter three. Chapter four offers a detailed description of the qualitative data collection and analysis. Results are presented in chapter five, which are subsequently compared to previous research in the discussion section. The final chapter concludes with implications and propositions for future research.

2 Food-scanner technology

Food-scanners are miniaturized and portable NIR spectrometers. NIR spectroscopy is a type of vibrational spectroscopy which depends on the stimulation of molecular vibration using infrared light in the wavelength range 750-2500 nm. The interaction of NIR electromagnetic waves with C-H, N-H, O-H or S-H molecular bonds of the samples constituents leads to spectra which is captured by an optical sensor (Pasquini, 2003). By applying multivariate statistical analysis, these spectra can be correlated with the trait of interest, for example sugar content, dry matter or firmness of fruit. Therefore NIR spectroscopy can be used to acquire quantitative and qualitative information from a sample by determining the physical and chemical composition of produce in a rapid and non-destructive way.

The latest innovations in this field of technology include smartphone-based and portable food-scanners like TellSpec Enterprise Scanner (TellSpec Inc.), SCiO (Consumer Physics) and FoodScanner (Spectral Engines Oy) (Rateni, Dario, & Cavallo, 2017). These wireless sensors can be operated via smartphone or tablet and use sample libraries and cloud-based prediction models to identify contents like fat, sugar, starch, moisture, protein and total energy in real-time (Consumer Physics, 2017a; Spectral Engines Oy, 2018). The operating principle of food-scanners is as follows (Figure 1): After illuminating the produce with NIR radiation from the light source (1) the food-scanner receives the spectrum through its detector (2). This spectrum is forwarded to a mobile device (e.g., smartphone, tablet) via Bluetooth (3), where a mobile application sends the spectrum via WLAN to a cloud database (4). Here, previously created sample libraries are used to analyze the spectrum using advanced algorithms (5). Finally, the result from the cloud database is sent to the mobile device and illustrated for the user (6). Some companies are already promoting mobile applications for end-consumers for the determination of sugar content in fruit (Consumer Physics, 2017b) or testing of fresh fruit for quality, ripeness and flavor (TellSpec Inc., 2018).

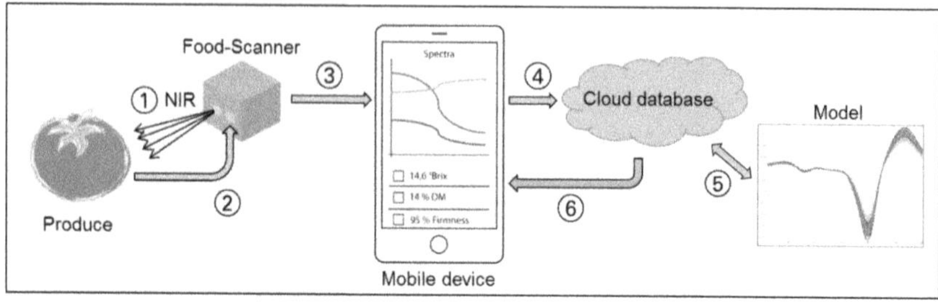

Figure 1: Operating principle of miniaturized food-scanners

An examination of various studies addressing the topic of portable NIR spectrometers shows that there are some advantages compared to traditional destructive and time-consuming quality control methods (dos Santos et al., 2013). Initial studies of the device SCiO showed acceptable results in predicting dry matter of apples and kiwifruit (Kaur, Künnemeyer, & McGlone, 2017) as well as predicting sugar content in kiwifruit and classifying feijoa according to maturity and 'Hass' avocado according to ripening stage (Li, Qian, Shi, Medlicott, & East, 2018). A recent study illustrated that these portable NIR sensors can be used for the prediction of shelf-life of stored tomatoes by predicting firmness and modeling the decline of firmness over storage time (Goisser, Krause, Fernandes, & Mempel, 2019).

3 Literature review

Originally some companies promoted portable food-scanners as tools for end-consumers, allowing real-time, portable analysis, assuring food safety and food security (TellSpec Inc., 2018) and aiming to be the consumer's "sixth sense" (Consumer Physics, 2017b). Since the use by end-consumers is an emerging field of application for food-scanners, experts have started discussing opportunities and threats through consumers using these food-testing devices for produce. According to these experts, main challenges arise in the scope of application and the interpretation of results. Lack of understanding and mistakes in handling by consumers as well as potential imprecise models can lead to false results. Additionally, sampling performed by non-expert users can result in adulterated measurements due to contaminations of samples (Rateni et al., 2017). Therefore, handling by consumers is a challenge, since incorrectly performed measurements can cause severe brand damage for producers and retail markets due to negative social media exposure and unnecessary food waste and recalls (Popping & Bourdichon, 2018). To mitigate these risks and help consumers better understand and interpret results, guidelines for consumer device manufacturers were developed (Popping et al., 2018). A first examination of smartphone-based diagnostic platforms by Rateni et al. (2017) described a user-friendly design of commercial systems.

Furthermore, the authors report instructions attached to these new devices to assist users during calibration and sampling procedure, indicating that measures against incorrect operation as described by Popping et al. (2018) are already taken adequately into account by device manufacturers.

Food-scanner as innovation to the FSC

A broad definition provided by Andersson, Lindgren, and Henfridsson (2008) describes innovation as “new application of knowledge, methods, and technologies that leverage an organization’s competitiveness” (p. 21). Based on the fact that portable food-scanners were originally designed for end-consumers, the implementation of food-scanners along prior steps of the FSC constitutes a “new application of technology”. First scientific studies show promising results regarding the prediction of internal quality as well as maturity of fruit, allowing a fast and non-destructive quality measurement. Therefore, the application of portable food-scanners as tools for quality measurement could leverage the competitiveness of an organization, since time-consuming and destructive measurements could be avoided.

Following Tidd's (2006) approach, the implementation of portable food-scanners as tools for quality assessment along the FSC can be classified as a process innovation within the FSC. According to Tidd (2006), opportunities for innovation can present themselves due to different triggers of discontinuity (e.g., new markets, new technologies) within a set of rules. These opportunities can be overlooked by established players because they are beyond the usual focus of attention. Also, the convergence and maturing of established technological streams, which in combination could offer a benefit in product or process technology, can be underestimated. The set of rules in the context of quality management are traditional methods of quality measurement, such as the determination of sugar content with refractometers and measuring firmness with penetrometers. Furthermore, the miniaturization of NIR spectroscopy and application as a new form of technology can be considered as a trigger, providing various opportunities in quality control processes. Actors along the FSC, which can be considered key stakeholders of this new technology, could potentially benefit from this innovation in measurement in the process of quality control.

Kim, Kumar, and Kumar (2012) divide innovation into five different types: incremental product, incremental process, radical product, radical process, and administrative. First, innovation is categorized as either administrative or technological innovation. Depending on the level of change, the target customer or market and the level of risk, technological innovation can be further categorized as incremental or radical innovation. Furthermore, the innovation subject refers to either a product or a process. Since the application of portable food-scanners introduces a new element for task specifications and workflow mechanisms, food-scanners can be considered as a radical process innovation in the FCS, even though for their producers they constitute a radical product innovation.

Sunding and Zilberman (2001), who examined the agricultural innovation process, give additional examples how innovations can be categorized and distinguished. Furthermore, they describe different ways of the generation of innovations and illustrate the process of adoption of new technologies. Referring to the "Cochrane treadmill" (Cochrane, 1979) and additional studies performed by Kislev and Shchori-Bachrach (1973), Sunding and Zilberman (2001) highlight the subgroup of the farming population called early adopters or early innovators. These farmers, who are the first to adopt a new technology, are able to benefit from technological change and the resulting profit. As further elaborated by Epperson and Estes (1999), early adopters within a technology-driven industry such as the fruit and vegetable industry are able to realize gains in productivity, higher than average season prices and enhanced market access.

With respect to reports on recent innovations in the field of horticulture there is high interest in new methods and technologies for postharvest produce treatment, e.g., ultrasound treatment of fruit to extend shelf life (Aday, Temizkan, Büyükcan, & Caner, 2013), UV-radiation of fresh-cut tropical fruit to enhance health promoting compounds such as phenols and flavonoids (Alothman, Bhat, & Karim, 2009), new developments in controlled atmosphere technology for storing organic produce (Prange, DeLong, Daniels-Lake, & Harrison, 2005), as well as new forms of packaging which allow the extension of shelf life and quality control of produce (Wyrwa & Barska, 2017).

Drivers and barriers for adoption of innovation

Most technological innovations follow the same pattern of organizational diffusion process (Hazen, Overstreet, & Cegielski, 2012). Rogers (2003) distinguished among five stages in this process, where the decision to adopt an innovation divides activities of initiation (gathering of information, conceptualization) and implementation (events, actions and decisions to utilize the innovation). Adoption therefore is the crucial step in realizing hitherto notional intentions. Within the literature, a number of drivers and barriers for innovation adoption have been identified. The following literature review focuses on drivers and barriers of adoption within the field of agriculture and horticulture and takes different steps along the supply chain into account.

By studying Irish farmers and clustering them into the two groups, innovators and adoption-averse farmers, Läpple, Renwick, and Thorne (2015) found a small but positive relation between farm size and innovation behavior, where innovators managed larger farms compared to adoption-averse farmers. The evaluation of farmer characteristics showed that younger farmers are less risk averse than older farmers. Additionally, the fact that farmers had completed agricultural education was positively correlated with innovation, concluding that agricultural education increases farmer's awareness of available innovations. Furthermore, the authors conclude that farmers with higher educational attainment might be more effective in processing new information. Pierpaoli, Carli, Pignatti, and Canavari (2013) studied adoption drivers for precision agriculture technologies and found that non-adopters do not have sufficient skills and competence in managing these technologies, which is in line with conclusions of Läpple et al. (2015). Furthermore, both studies conclude a positive relation between access to credit as well as financial resources and innovation. Dewar and Dutton (1986) investigated the adoption of radical and incremental innovations. According to the authors, the adoption of radical innovations is highly affected by knowledge resources and size of the company, corresponding with findings of the aforementioned studies. Additionally, Pierpaoli et al. (2013) emphasized the importance of in-field demonstrations, support services and free trials as measures related to the use of new technologies, since these practices help farmers to perceive the use of a technology as easy. Aspects such as ease of use and usefulness are central aspects for the adoption of a new technology. The authors also highlight the need for an intrinsic simplicity of a new technology to avoid incompatibilities among different tools as well as difficulties in using different technological devices simultaneously. Kafetzopoulos and Skalkos (2019) examined the innovation capability of Greek agri-food firms and identified quality orientation and process management as the two most important innovation drivers. According to the authors' findings, quality orientation within a company creates a productive environment for innovation development, whereas effectively managed processes lead to concentration on important issues as well as avoidance of irrelevant activities.

Soderlund, Williams, and Mulligan (2008) investigated the adoption of assurance systems as an innovation in different agri-food value chains. According to their study, numerous strong drivers were successful in embedding the innovation in everyday practices. In contrast, just one driver led to minimalist behavior. As an example, producers in the cherry chain only adopted the new assurance system with the sole purpose of gaining access to supermarkets. Furthermore, the authors found that high complexity and cost of this innovation represented a major barrier to the adoption by small cherry producers. Additionally, manufacturers, processors and packers within each agri-food value chain performed as hubs and set the standards for adoption of various forms of assurance within the chain. These findings are in line with a study from Fortuin and Omta (2009) who observed a very high pressure on food-processing companies from their buyers within the retail sector. This unequal distribution of power within the value chain was identified as a powerful driver for the engagement in innovation.

With respect to the acceptance of innovations within a company, Talukder (2019) examined drivers and barriers of technological innovations by individual employees. In this study organizational support was found to be an important factor for an uncomplicated adoption process. Therefore, management needs to provide the necessary administrative and technical support such as spreading of information to and training for employees. Additionally, motivation of employees such as proper incentives by means of recognition, job security, and increased autonomy were found to influence employees' willingness to try and finally practice a technological innovation. As for innovation barriers, too many innovative features as well as too frequent changes in these features were identified. Pantano (2014) studied the retail industry and identified three main factors which influence innovation. On the one hand, consumers are demanding innovation at the point of sale with respect to entertaining elements, which increase the quality of the shopping experience, as well as supporting tools, which enhance their empowerment. As further elaborated by Pantano and Viassone (2014), especially young consumers are interested in new technologies that allow a support of their purchasing decision. As examples, interactive tools for searching, comparing and tasting products are highlighted. Attention should be paid to privacy of consumer data, since concerns could reduce users' intentions to use these tools. On the other hand, Pantano (2014) identified the availability of new software tools which allow the understanding of market trends and prediction of consumer behavior as important innovation drivers. Using these tools, customized future advertising and selling strategies can be deployed to maintain competitive advantage. As for a third factor, the acceptance of frontline employees' of these new technologies seems to be of high importance, but current literature is still scarce in that regard.

According to Douthwaite et al. (2001), the adoption rate of a new technology is subject to the key stakeholders, the direct beneficiaries from an innovation, as well as the researchers who provide scientific background and were involved in the development of the prototypes. A working partnership between these two groups allows researchers to impart scientific knowledge to stakeholders as well as learn about the performance of the innovation in work environments. Using promising innovations in the context of agricultural technologies as examples, Douthwaite et al. (2001) furthermore illustrate that modifications made by the key stakeholders can lead to either beneficial or detrimental changes in the fitness of the new technology. Therefore, stakeholders and researchers need to work together to prevent that knowledge gaps result in mistakes, which impede the fitness of the innovation. Especially at an early release of a new technology combining key stakeholders' and researchers' involvement is crucial for generating rapid improvements.

4 Materials and Methods

In order to explore the research question in-depth, the present study employed a qualitative research approach. A qualitative approach is suitable when new or not well-known research issues are investigated or the research is aimed at future issues (Bitsch, 2005). Previous studies related to portable food-scanners focused on prediction-accuracy of selected quality parameters like dry matter and sugar content (Kaur et al., 2017; Li et al., 2018). The perception and evaluation of FSC actors regarding the implementation of this technology as a new way of measuring quality are yet unknown, therefore a qualitative approach is appropriate. Furthermore, quality management differs from company to company due to internal organizational structures, consequently semi-structured interviews were deemed suitable for data collection.

In September 2018 the German Fruit & Vegetable Congress in Dusseldorf as one of the most important congresses within the German fruit and vegetable supply chain was used to make contact with actors along the supply chain as potential research participants. Since attendants of this congress hold key positions along the German FSC and deal with quality management on a daily basis they were asked to participate in interviews and to suggest potential candidates for additional interviews. As a result, interviewees from two producer cooperatives, three wholesale companies, one logistics enterprise and one food distribution center could be recruited. With the objective of describing the whole FSC from producer to food retail market, interviewees from production companies and food retail markets were recruited through the researchers' personal contacts. In total thirteen semi-structured interviews were conducted with actors on different steps of the FSC from companies in different parts of Germany from fall 2018 to spring 2019 (Table 1).

All interviews were conducted by the first author. Four interviews were conducted face-to-face at the interviewees' companies, in quiet and neutral rooms, e.g., conference or break rooms. The remaining nine interviews were conducted via phone. The interviews lasted between 25 to 60 minutes. Each interview started with the current quality measurement at the interviewees' respective companies. Since all interviewees were familiar with these topics from their daily work, these questions also served as icebreakers, and allowed an easy start into the discussion. The following questions addressed potential applications as well as concerns and preferences of supply chain actors regarding the implementation of the technology. Furthermore, requirements for the practical use in their operations, opinions about possible areas of application outside the interviewees' companies and opinions about the potential consequences of end-consumer use of these devices were explored. An interview guide was used and all topics were addressed over the course of the interview following the flow of conversation. Before conducting the first interview for the study, the interview guide was tested and adjusted for comprehensibility. Adjustments were made by rearranging questions to improve interview flow and rewording questions to facilitate interviewees' comprehension of topics. Questions addressing the motivation of supply chain actors were added to the interview guide halfway through the process of data collection, since this topic was brought up and deemed important. The procedure of adding questions and therefore extending the focus of a research project is in line with qualitative research procedures and indicates a maturing research process (Bitsch, 2005).

Table 1: Interviewees and their background

Company position along the FSC	Interviewee	Duty
Fruit production	Employee	Cultivation and direct selling of fruit
Fruit production	Owner-manager	Cultivation and direct selling of fruit
Vegetable production	Manager	Cultivation and direct selling of vegetables
Producer cooperative	Sales manager	Managerial and administrative duties
Producer cooperative	Quality manager	Managerial and administrative duties
Wholesale fruit and vegetables	Trade manager	Managerial and administrative duties
Wholesale fruit and vegetables	Regional manager	Managerial and administrative duties
Wholesale fruit and vegetables	Quality manager	Managerial and administrative duties
Logistics	Project manager	Managerial and administrative duties
Food distribution center	Team manager fruit and vegetables	Managerial and administrative duties
Food retail market	Branch manager	Orders produce and monitors quality
Food retail market	Department manager fruit and vegetables	Orders, sorts and shelves produce
Food retail market	Department manager fruit and vegetables	Orders, sorts and shelves produce

All interviews were audio-recorded and transcribed verbatim. Content off-topic was omitted during transcription (Halcomb & Davidson, 2006). Since the present study focused on the content of the interviews, the simple transcript method (Dresing, Pehl, and Schmieder, 2015) was used. Therefore, colloquial language and dialect were adjusted to standard German language. Qualitative content analysis using the Atlas.ti software (version 8.2.32.0) was applied to structure the results, utilizing coding and the establishment of categories. In the process of open coding, sections of the text were labeled and assigned with the main thought behind each section. Afterwards categories were established by grouping codes together according to their meaning and the relationships between them. A category with three codes, respective definitions and examples of interview excerpts is provided as illustration of the analysis process (Table 2).

Table 2: Codes for the category "motivation for the application of portable food-scanners" with excerpts from interviews

Code	Interview excerpt
Intrinsic motivation Interviewees' motivation is driven by intrinsic forces, e.g. to motivate employees, be innovative and distinguish oneself from competitors due to better quality	"Exactly, that is absolutely the case, to know the newest technology in our immediate environment, and then I have to decide for myself, yes that could be of interest for us, because we get better results, because we save time, because whatever. This motivation is absolutely there." (Quality manager of a wholesale company of fruit and vegetables)
Extrinsic motivation Motivation to apply food-scanners comes from outside the interviewees' own company	"At the very moment we have a requirement, from trading companies, from our customers, we have to deploy that, we have to produce results. And in order to produce reliable results I have to be ready in this moment to say okay, we will do it." (Quality manager of a producer cooperative)
Non-existent motivation At the moment there is no need and motivation for the application of food-scanners	"And at the moment, I don't see the need. I say now perhaps, as mentioned earlier, I am in the lucky position to have employees who can then also say: that is all right, this is not all right. (Owner-manager of a fruit production company)

5 Results

Results are structured into five parts. The first part describes the status quo of quality assessment along the supply chain, including the experiences of interviewees along the FSC in their day-to-day work. The second part analyzes their preferences and concerns regarding the implementation of portable food-scanners for quality control. The third part presents the requirements food-scanners have to fulfill to be of practical use in quality management. The fourth part highlights potential applications of food-scanners along the FSC. The fifth part characterizes different patterns of motivations of supply chain actors for implementing portable food-scanners as tools for quality control.

Current practice of quality assessment along the FSC

When asked about their daily routine in the context of quality control of fresh produce, actors described practices at their respective companies to ensure high quality. Since differences in these practices with regard to different supply chain steps (production, trade, retail) could be identified, the description is divided into these parts. According to interviewees from fruit production companies, quality, maturity as well as harvest date of fruit is often monitored via cultivation consultants. These consultants provide suggestions whether harvest time is reached. Therefore, quality assessment at that stage is mostly limited to optical inspection of fruit in the orchard. Internal quality parameters such as firmness and sweetness are solely examined through subjective testing, e.g., hand-squeezing and degustation in the field. Furthermore, as experience is established over time, working in the business is paramount for producers, rendering additional testing obsolete.

“We don’t have a device or anything. Of course, we are going in [the orchard], I would say, we take a look, we taste, and over time we developed a feeling, kind of an experience when something is ready for harvest” (Employee in fruit production, male, 20-30 years old).

To sell fruit to central markets and reach a high price marketing and commercial quality control standards have to be fulfilled, e.g., a specific degree of fruit coloring and specific fruit size. Internal quality parameters like sugar content are often not specified by central markets. Actors handling fruit along the FSC (producer cooperatives, wholesale, and fruit distribution centers) described a combination of external and internal quality parameters regarding incoming produce. In a first step, fruit is visually inspected and conformity of sizes, colors, weights and storage temperatures are verified, sometimes in combination with hand-squeezing to test fruit firmness. For this assessment of quality, the know-how and experience of the staff is important and dominated by visual impressions. In a second step, quality parameters related to internal quality are tested, e.g., sugar content using refractometers or fruit firmness using penetrometers (Figure 3). One interviewee stated that in addition to these objective measurements a degustation of fruit was implemented to verify fruit quality and taste.

“There is always a degustation, though. So, we slice [the fruit] and take a bite. But then as subjective evaluation, is it well, is it not well” (Quality manager at wholesale fruit and vegetables, male, 40-50 years old).

On the one hand these tests are performed to make sure standards regarding marketing and commercial quality control (UNECE, 2017) are met, on the other hand to fulfill customer-specific standards which can be stricter than the aforementioned standards. Some of these tests are described as elaborate and costly, e.g., the determination of sugar content or the brix/acid ratio, and criticized for a lack of reproducibility. Other tests like the determination of dry matter are characterized as difficult, since employees require special knowledge, which results in outsourcing of analyses to laboratories.

According to actors at retail markets, quality evaluation of fruit and vegetables is divided into two sections, first, control of marketing and commercial quality control standards (e.g., origin, grade) and, second, assessment of fruit quality during sorting and shelving of produce. This procedure of quality control is solely based on visual inspection and haptic testing of fruit firmness and described as tedious and not particularly hygienic work which sometimes leads to defects being overlooked due to monotonous and repetitive work. Experience of the retail staff responsible for sorting and shelving is paramount, since criteria for rejecting produce are sometimes vaguely phrased.

"Well the company by itself specifies that they say fruit, which oneself would no longer buy. So it is relatively vague, fruit and vegetables which oneself would no longer buy" (Branch manager at food retail market, male, 30-40 years old).

Preferences and concerns regarding the implementation of food-scanners

When interviewees were asked where they see the main advantages of portable food-scanners several aspects were identified. Compared to traditional methods of quality evaluation, the non-destructive nature of food-scanners posed an important advantage, making destructive tests like refractometer analysis obsolete. Produce must no longer be touched, allowing a more hygienic workflow. Also, the speed of measurement of internal quality parameters was deemed advantageous. The time saved in comparison to destructive measurements could be used more efficiently, e.g., by increasing the number of inspections at each arrival of produce. Additionally, laboratory analyses could be reduced to a minimum. A rapid, nondestructive and laboratory-independent measurement would save money for companies.

According to several interviewees, portable food-scanners could provide an opportunity for objectifying current measurement methods, which are mainly visual and based on staff's experience. Consequently, food-scanners could replace subjective grading. As a result, incoming employees could assist earlier in produce control with portable food-scanners as support. As mentioned by various interviewees, food-scanners could further be used as additional decision-support tools to accept or refuse produce at incoming goods control. For instance, fruit and vegetables could be tested via conventional methods (e.g., visual inspection) and food-scanners added to give information about internal quality attributes, indicating if produce passes all requirements or a complaint has to be filed. Consequently, feedback can be directed to suppliers, allowing communication of internal quality along the FSC.

Some actors perceive a downside of the fact that internal quality measurements could be easily available by subsequent purchasers in the FSC. According to these critical voices, additional pressure for producers could arise due to additional quality requirements from fruit wholesalers, where specified scan results have to be met to be accepted as good quality. Critical attention is also paid to the accuracy of the new devices. To avoid discrepancies between scan results at different levels of the supply chain, the transferability of predictions from different devices has to be guaranteed. Informative and accurate predictions are required to allow long-term utilization of these devices and prevent frustration.

The application of portable food-scanners by end-consumers is viewed critically. Some interviewees hold the opinion that most consumers will not purchase an extra device like a food-scanner for testing food, but will more likely give it a try when implemented in smartphones. Even then, most supply chain actors do not perceive portable food-scanners as devices for the general public, but rather for small consumer-groups with special interests in health and diets. On the other side, some actors see potential in these devices as an additional incentive and decision-support for consumers during purchasing decisions, especially for a young and health-conscious target group.

Another aspect critically discussed is the impact of the application of food-scanners on food waste. As a result of the above mentioned advantages like fast quality evaluation at incoming goods control, more and precise quality measurements could be required, leading to a higher rejection rate and thus to more food waste. Furthermore, additional food waste could emerge at retail markets due to end-consumers being more selective in their choice. However, the increase is not expected to be large compared to current amounts.

Requirements of food-scanners to be of practical use in day-to-day processes of quality control

An important requirement of portable food-scanners to allow the application in day-to-day quality control processes according to various interviewees are investment costs. As stated by multiple actors along the supply chain, devices which cost several thousand euros are unlikely to be purchased. Besides investment costs, the potential revenue of food-scanner applications also is of high importance. According to interviewees the application of food-scanners would be more realistic if additional benefits such as a higher number and more accurate test results as well as time-saving due to faster measurements could be achieved.

Another important aspect, mentioned by many interviewees, is the fact that the reference models and the database used as well as the predictions of quality have to be accurate and reliable. Additionally, it is important, whether there will be different prediction models for each variety of produce or one global model which allows the measurement of all varieties of this produce. Since the technique and the way of operating with these devices would be a radical process innovation compared to traditional quality measurement procedures for all interviewees, almost all stated the importance of building confidence as an important step in adopting these devices in day-to-day routine measurements. According to interviewees, results would be cross-checked with traditional measurement methods like refractometer or penetrometer tests during the introduction phase to generate experience. After accumulating evidence for the reliability of food-scanner predictions, these cross-checks would be terminated and the devices used independently. Several interviewees highlighted the importance of media reports, which describe and verify a high accuracy of measurement of portable food-scanners in order to be acknowledged as adequate tools for quality assessment. Regarding the operability of food-scanners two factors are important to allow application in day-to-day quality control processes. On the one hand, handling of these devices by employees has to be as easy as possible to allow a fast and simple workflow, as stated by one interviewee.

"Those must be easy to use devices, which allow the result by a simple push of a button" (Team manager fruit and vegetables at food distribution center, male, 40-50 years old).

On the other hand, display of measured values has to be arranged in a way to allow easy interpretation. As suggested by several interviewees, a traffic light food labelling system with colors such as green, yellow and red could serve as an indicator to support decisions for employees at incoming goods control. With regard to the utilization by end-consumers, results should be displayed in a comprehensible way without technical terms to avoid confusion and unintended waste.

Another requirement besides operability is a certain convenience in handling and robustness of the devices (Figure 2). Robustness was mentioned by both interviewees from fruit production with regard to the nature of work in orchards, so that devices withstand falling to the ground during utilization in the field. Due to rough environmental conditions in fruit wholesale, a similar robustness is required at this step of the supply chain, as mentioned by one interviewee.

"The device now has to withstand the environment in our warehouse, at our produce arrival, so it must survive falling down. It will be touched by wet, sticky hands. It has to withstand the cold, the high humidity, so these environmental conditions at our cold storage" (Quality manager at wholesale fruit and vegetables, male, 40-50 years old).

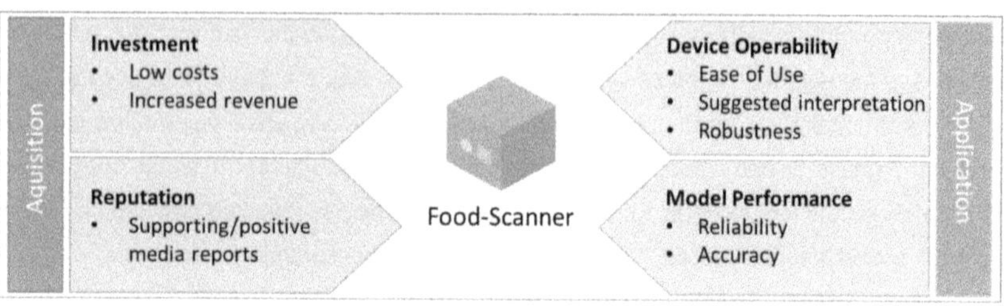

Figure 2: Requirements of food-scanner to be of practical use as stated by interviewees

Potential applications of food-scanners along the FSC

Interviewees perceived different fields of possible applications of portable food-scanners along the FSC. For instance, sorting produce as well as the determination of maturity and ripeness, especially in the context of produce new to the assortment, are viewed as promising applications. Furthermore, product-specific applications such as the detection of internal fruit damage or prediction of taste patterns such as sweetness of fruit will make food-scanners attractive to use, allowing a leap forward in quality assessment, as illustrated by one interviewee.

"Principally one topic, where I currently don't know if such a device is capable of, would be testing of internal browning for avocados, discovering internal deterioration for pineapple or detecting internal browning for mango. [...] So, if something like this would be possible that would imply a big leap forward for us in [quality] control" (Quality manager at wholesale fruit and vegetables, male, 40-50 years old).

Additionally, the potential of implementing food-scanners is considered promising along the whole FSC. At production, portable food-scanners could be used to determine optimum harvest time, ripeness and maturity of fruit, whereas wholesalers and retailers could profit from food-scanners by using them as tools for fast and objective quality assessment of incoming and outgoing produce. Additionally, these devices could be useful as decision support tools in retail markets to promote the internal quality of produce to a health-conscious target group.

Motivation of actors along the supply chain for implementing food-scanners as a tool for quality control

Although not specifically addressed by interview questions in the beginning, different patterns of motivation for the implementation of food-scanners as tools for quality assessment could be distinguished. In order to further evaluate this topic, questions addressing the motivation of supply chain actors were added in subsequent interviews. Some interviewees' motivation seems to be driven intrinsically. According to one interviewee, food-scanners could be helpful for profiling the quality of fresh produce up to the next step of the supply chain as well as distinguishing good from mediocre quality:

"As I have previously mentioned our aspiration is that we produce the highest quality, and there I don't have to hide and I would not be afraid of [the application of food-scanners], because when I say that, I want to have it and do it that way. [...] But nevertheless there are black sheep [among producers] and I believe that especially those could be taken out of trading" (Employee at fruit production, male, 20-30 years old).

Furthermore, this sort of profiling of quality also seems of interest for retailers, where food-scanners could be implemented in advertising quality and new varieties of produce to customers. Compared to traditional quality control, this could provide additional value, as described by one interviewee:

"And so one would have this device as a tool. Then I could scan my tangerines in the morning and it tells me they are sweet and then I can attach an extra note and say 'today extra great'" (Department manager fruit and vegetables at retail market, female, 40-50 years old).

Besides profiling quality, aspiration for new technologies as innovation was identified as a form of intrinsic motivation of utilizing food-scanners. According to various interviewees it is important to stay up-to-date with technology. Given the above mentioned condition that prediction accuracy were guaranteed and devices passed cross-checks, application of these devices in daily work routine would not pose a problem. Furthermore, implementation of portable food-scanners would also be exciting and comfortable for employees compared to traditional destructive measurements.

In contrast to these intrinsic patterns of motivation, the second pattern can be viewed as extrinsic motivation. According to some interviewees, the implementation of food-scanners would have to be a requirement demanded by trade companies in order to continue business relationships. Since at the moment there oftentimes is no requirement in providing information about internal quality, companies have no benefit in performing and providing these measurement results. Additionally, interviewees described an already high and stressful workload, which does not allow the experimental implementation of a new and uncertain technology. Until buyers demand consequent internal quality measurements, an application in companies of these interviewees seems rather unlikely.

As for a third pattern, one interviewee seemed to have no motivation at the moment to implement food-scanners in his respective business in vegetable production. According to the interviewee's statements, he has no need to apply this technology since his employees are very well trained and know to distinguish good from bad quality from years of experience. Although being conservative in introducing this technology in his own company there is understanding for applying these devices in companies further down the supply chain where there is a higher percentage of untrained personnel and a larger variety of produce. There, portable food-scanners could constitute a technical support besides training courses to distinguish different qualities.

6 Discussion and conclusions

The study showed that currently there are different practices of quality control for fresh produce along the supply chain, varying between production, wholesale and retail companies. Quality determination is often highly dependent on trained and experienced staff performing subjective and mostly visual examination of produce. In addition, wholesalers often follow established protocols to ensure quality and conformity with standards. Various interviewees described portable food-scanners as tools for objective quality measurement. Therefore, these devices could help to overcome existing discrepancies in quality assessments along the FSC by establishing a uniform measurement method. These findings correspond with Abbott (1999) who stated that instrumental measurements are often preferred over sensory evaluation since they allow reduction of variation between individuals, offer a higher precision and provide a standardized language among researchers, industry and consumers. Additionally, potential applications which could arise through the implementation of food-scanners were described by actors along the FSC (Figure 3).

Results indicate that there are several perceived advantages of food-scanners compared to traditional methods of quality evaluation that make these devices attractive to use. Aspects such as lower costs due to fewer laboratory costs, non-destructive and rapid measurements are of high importance as identified in this study and underline the advantages of NIR spectroscopy mentioned in the literature (dos Santos et al., 2013). Since the current study focused on day-to-day application along the FSC, new aspects such as the application by untrained employees and utilization as decision-support tools for incoming goods were identified as additional advantages. However, concerns of some supply chain actors due to the possibility of additional quality requirements from fruit wholesalers, resulting in further pressure for producers, were identified. These results confirm findings that small producers can face considerable challenges meeting the requirements of retail chains as described by Boselie, Henson, and Weatherspoon (2003).

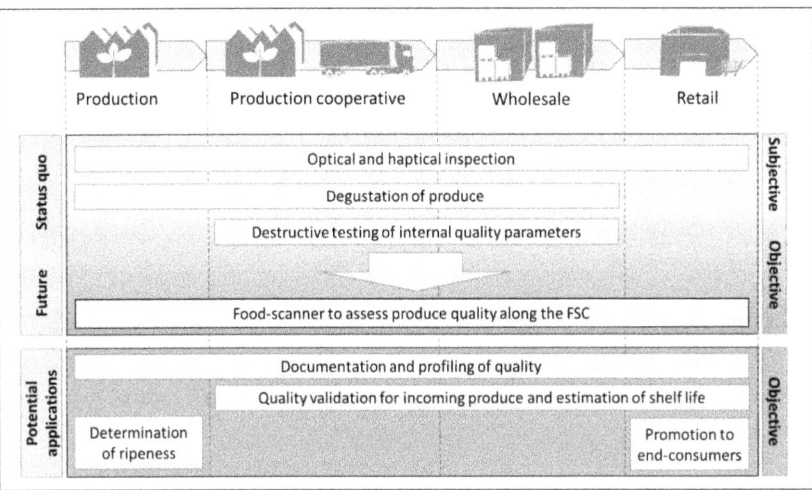

Figure 3: Status quo of quality assessment along the FSC and potential applications of food-scanners through their implementation along the FSC

The prior literature on implications of consumer testing devices for the industry is limited to the need of developing guidelines to help consumers interpret and understand test results in order to allow a save consumption of products as well as reduce the risk of recalls due to false measurement results (Popping et al., 2018; Popping & Bourdichon, 2018). With respect to consumer's complex decision-making process (Kaine, 2004), portable food-scanners could assist users during the step of information processing by searching, screening and gathering of information, therefore simplifying product evaluation and purchase decision. Onwezen and Bartels (2011) identified a health-oriented consumer group, which places value on healthiness, taste and safety aspects of food. According to interviewees' statements, especially these health-conscious consumers could benefit from food-scanner supported decisions. Results of the present study confirm the need for an easy interpretation of results (Popping & Bourdichon, 2018). Furthermore, results indicate that there are additional aspects that have to be considered with regard to the implementation of portable food-scanners, such as the impact on food waste. Drivers and barriers of the adoption of new and smart technologies by consumers have been identified in several studies (Antioco & Kleijnen, 2010; Arts, Frambach, & Bijmolt, 2011; Joachim, Spieth, & Heidenreich, 2017; Mani & Chouk, 2016) and should be considered by food-scanner manufacturers, especially when a widespread incorporation of miniaturized NIR spectrometers into the hardware of future smartphones is considered.

Investment costs, high prediction accuracy, convenience in handling and robustness of food-scanners were identified in the present study as important requirements to allow the application in daily quality control processes. These critical requirements were also identified by dos Santos et al. (2013). The preference of supply chain actors of integrating food-scanners in existing systems of information technology posed an additional requirement identified in the current study. To date, food-scanner systems require the connection to a mobile application and database provided by food-scanner manufacturers (Figure 1). In order to enable the integration of these devices in existing systems of information technologies, supply chain actors and food-scanner manufacturers need to work together to avoid a decrease in fitness of the innovation (Douthwaite et al., 2001). Otherwise, a too high complexity of this new technology could constitute a barrier for the adoption of food-scanners, as similarly described by Soderlund et al. (2008) for the implementation of assurance systems by cherry farmers.

Trienekens, van Uffelen, Debaire, and Omta (2008) assessed innovation and performance in the fruit supply chain by analyzing Dutch apple growers and a fruit cooperative. Their findings indicate that on the cooperative level, the most important innovations are process innovations. Furthermore, the most important critical success factor for innovation in a cooperative is process superiority, which refers to a fully automatic first-in-first-out (FIFO) system. There is a constant drive for innovation with respect to logistics, storage and processing techniques. When the application of food-scanners becomes feasible in the near future and the prediction of shelf-life is possible, this could lead to a shift from current standards like first-in-first-out to new standard systems such as a "least shelf-life, first-out". Therefore, produce with longer shelf-life could be stored for a longer period of time at cooperatives and allow produce which shorter shelf-life a faster turnover, preventing food-waste. At this moment, this potential development is a hypothesis which should be investigated in future research.

Additionally, the interviews performed by Trienekens et al. (2008) revealed that an information system, e.g., specialized newspapers, magazines, fruit advisor recommendations and visits to research facilities, is crucial for Dutch apple growers to keep up-to-date with new and emerging technologies. Results from the present study confirm these findings, since several interviewees highlighted the importance of media reports about these new devices with respect to reliability and accuracy of quality predictions. Measures such as in-field demonstrations and free-trials (Pierpaoli et al., 2013) as well as professional development courses (Läpple et al., 2015) have been highlighted as opportunities to promote innovation by broadening farmers' horizons. Therefore, information events and advanced training sessions could be first actions in promoting the application of portable food-scanners along the whole FSC.

Farm size, farmer's age, the level of education and the availability of financial resources have been described in the literature (Dewar & Dutton, 1986; Läpple et al., 2015; Pierpaoli et al., 2013) as factors influencing innovation adoption, but due to the qualitative nature of this study, these aspects were not deemed representative. Statements from interviewees on the production level regarding the intrinsic motivation to produce high quality are in line with findings from Kafetzopoulos and Skalkos (2019) and can be considered a driver for the adoption of food-scanners as an innovation. According to statements by one interviewee from a producer cooperative, food-scanners would only be implemented if they were stipulated requirements from their buyers. This result is in line with findings from Fortuin and Omta (2009) who identified high pressure from buyers and unequal power distribution along the value chain as powerful driver for the adoption of innovations. It is important to note that for future applications of food-scanners there should be additional incentives for providing these measurements, e.g., financial compensation for high quality produce. Otherwise the sole driver of applying food-scanners in gaining market access could lead to minimalist behavior, as described for agri-food assurance systems by Soderlund et al. (2008). Interviewees in retail markets deemed portable food-scanners useful as decision support tools for end-consumers, since these devices could help promote fruit quality to a health-conscious target group. This aspect confirms results from Pantano (2014) identifying consumer demand for innovations as the main driver in the retail industry. Therefore, new technologies could serve as entertaining elements and supporting tools enhancing the shopping experience and increasing consumers' empowerment.

Magwaza et al. (2012) described the commercial utilization of NIR technology in pack-houses and fruit sorting lines since the mid-1990s. The determination of maturity and sorting of produce according to ripeness are desired by actors as potential applications of food-scanners along the FSC as highlighted in this study. A recent study by Li et al. (2018) indicated the possibility of using portable food-scanners for this type of application and could be feasible in the near future. The detection of internal fruit damage by applying NIR spectroscopy seems generally possible (Fu, Ying, Lu, & Xu, 2007). However, due to several advantages on the basis of the operating principle (e.g., applying transmittance compared to reflectance spectroscopy), internal defects are more likely to be detected using hyperspectral imaging (Ariana & Lu, 2010).

According to Banks, Maguire, and Tanner (2000), "fruit industries of the world are beginning to recognize the revolutionary potential of technologies for characterizing invisible aspects of product quality" (p. 294). As they further specified, these technologies could help to separate fruit based on invisible traits such as storage behavior, susceptibility to shrivel, and flavor. Recent studies by Goisser et al. (2019) on shelf life prediction of tomatoes and Jamshidi, Minaei, Mohajerani, and Ghassemian (2012) on taste characterization of oranges indicate that NIR could be a technology with the potential of detecting such invisible traits. As the present study has shown, actors along the FSC also see possible applications of food-scanners in the prediction of taste patterns such as sweetness of fruit as well as the detection of internal fruit damage. Therefore, food-scanners could lead to a radical process innovation within the field of quality assessment of fruit, since conventional and time-consuming measurement procedures could be replaced and reduced to a minimum. As further elaborated by Banks et al. (2000), these new technologies could enhance consumers' eating experience in eliminating the current "unpredictable lottery of satisfaction" (p. 294), which at the moment is mainly based on outward appearance of produce.

The present study highlighted preferences and concerns of actors along the FSC with respect to the application of portable food-scanners as additional tools for quality assessment. Since food-scanners are already commercially available and expected to be used more broadly along the supply chain of fresh produce soon, future work should focus on evaluating critical points like the impact on quality perception along the supply chain, guaranteeing a development of the new technology according to the goals of food-scanner manufacturers as well as stakeholders along the FSC. To evaluate possible consequences due to food-scanners being used by end-consumers, future work should critically evaluate the general readiness of end-consumers in applying these devices in everyday life. Furthermore, to facilitate the widespread application of food-scanners it has to be possible to interpret predictions by food-scanners by non-experts.

Acknowledgement

The project was supported by the Bavarian Ministry of Food, Agriculture and Forestry as part of the alliance "Wir retten Lebensmittel" [We Save Foodstuffs].

References

Abbott, J. A. (1999). Quality measurement of fruits and vegetables. Postharvest Biology and Technology, 15(3), 207–225. https://doi.org/10.1016/S0925-5214(98)00086-6

Aday, M. S., Temizkan, R., Büyükcan, M. B., & Caner, C. (2013). An innovative technique for extending shelf life of strawberry: Ultrasound. LWT - Food Science and Technology, 52(2), 93–101. https://doi.org/10.1016/j.lwt.2012.09.013

Alothman, M., Bhat, R., & Karim, A. A. (2009). UV radiation-induced changes of antioxidant capacity of fresh-cut tropical fruits. Innovative Food Science & Emerging Technologies, 10(4), 512–516. https://doi.org/10.1016/j.ifset.2009.03.004

Andersson, M., Lindgren, R., & Henfridsson, O. (2008). Architectural knowledge in inter-organizational IT innovation. The Journal of Strategic Information Systems, 17(1), 19–38. https://doi.org/10.1016/j.jsis.2008.01.002

Antioco, M., & Kleijnen, M. (2010). Consumer adoption of technological innovations: Effects of psychological and functional barriers in a lack of content versus a presence of content situation. European Journal of Marketing, 44(11/12), 1700–1724. https://doi.org/10.1108/03090561011079846

Ariana, D. P., & Lu, R. (2010). Evaluation of internal defect and surface color of whole pickles using hyperspectral imaging. Journal of Food Engineering, 96(4), 583–590. https://doi.org/10.1016/j.jfoodeng.2009.09.005

Arts, J. W.C., Frambach, R. T., & Bijmolt, T. H.A. (2011). Generalizations on consumer innovation adoption: A meta-analysis on drivers of intention and behavior. International Journal of Research in Marketing, 28(2), 134–144. https://doi.org/10.1016/j.ijresmar.2010.11.002

Banks, N. H., Maguire, K. M., & Tanner, D. J. (2000). Innovation in Postharvest Handling Systems. Journal of Agricultural Engineering Research, 76(3), 285–295. https://doi.org/10.1006/jaer.2000.0579

Bitsch, V. (2005). Qualitative Research: A Grounded Theory Example and Evaluation Criteria. Journal of Agribusiness, 23(1), 75–91. https://doi.org/10.22004/ag.econ.59612

Boselie, D., Henson, S., & Weatherspoon, D. (2003). Supermarket Procurement Practices in Developing Countries: Redefining the Roles of the Public and Private Sectors. American Journal of Agricultural Economics, 85(5), 1155–1161. https://doi.org/10.1111/j.0092-5853.2003.00522.x

Cochrane, W. W. (1979). The development of American agriculture: A historical analysis. U of Minnesota Press.

Consumer Physics (2017b). It's sci-fi at your fingertips: Scan physical objects and uncover a world the eye cannot see. Available at https://www.consumerphysics.com/scio-for-consumers/ (accessed on February 27, 2020)

Consumer Physics (2017a). Solutions: Food & beverage quality control. Available at https://www.consumerphysics.com/business/solutions/ (accessed on February 27, 2020)

Dewar, R. D., & Dutton, J. E. (1986). The Adoption of Radical and Incremental Innovations: An Empirical Analysis. Management Science, 32(11), 1422–1433. https://doi.org/10.1287/mnsc.32.11.1422

Dos Santos, C. A. T., Lopo, M., Páscoa, R. N. M. J., & Lopes, J. A. (2013). A review on the applications of portable near-infrared spectrometers in the agro-food industry. Applied Spectroscopy, 67(11), 1215–1233. https://doi.org/10.1366/13-07228

Douthwaite, B., Keatinge, J. D. H., & Park, J. R. (2001). Why promising technologies fail: the neglected role of user innovation during adoption. Research Policy, 30, 819–836. https://doi.org/10.1016/S0048-7333(00)00124-4

Dresing, T., Pehl, T., & Schmieder, C. (2015). Manual (on) Transcription. Transcription Conventions. Software Guides and Practical Hints for Qualitative Researchers. 3rd English Edition. Available at http://www.audiotranskription.de/english/ (accessed on February 27, 2020)

Epperson, J. E., & Estes, E. A. (1999). Fruit and Vegetable Supply-Chain Management, Innovations, and Competitiveness: Cooperative Regional Research Project S-222. Journal of Food Distribution Research, 30(856-2016-56977), 38–43. https://doi.org/10.22004/ag.econ.27221

Fortuin, F. T.J.M., & Omta, S.W.F. (2009). Innovation drivers and barriers in food processing. British Food Journal, 111(8), 839–851. https://doi.org/10.1108/00070700910980955

Fu, X., Ying, Y., Lu, H., & Xu, H. (2007). Comparison of diffuse reflectance and transmission mode of visible-near infrared spectroscopy for detecting brown heart of pear. Journal of Food Engineering, 83(3), 317–323. https://doi.org/10.1016/j.jfoodeng.2007.02.041

Goisser, S., Krause, J., Fernandes, M., & Mempel, H. (2019). Determination of tomato quality attribute using portable NIR-sensors. OCM 2019 - Optical Characterization of Materials: Conference Proceedings. Advance online publication. https://doi.org/10.5445/IR/1000092314

Halcomb, E. J., & Davidson, P. M. (2006). Is verbatim transcription of interview data always necessary? Applied Nursing Research : ANR, 19(1), 38–42. https://doi.org/10.1016/j.apnr.2005.06.001

Hazen, B. T., Overstreet, R. E., & Cegielski, C. G. (2012). Supply chain innovation diffusion: going beyond adoption. The International Journal of Logistics Management, 23(1), 119–134. https://doi.org/10.1108/09574091211226957

Jamshidi, B., Minaei, S., Mohajerani, E., & Ghassemian, H. (2012). Reflectance Vis/NIR spectroscopy for nondestructive taste characterization of Valencia oranges. Computers and Electronics in Agriculture, 85, 64–69. https://doi.org/10.1016/j.compag.2012.03.008

Joachim, V., Spieth, P., & Heidenreich, S. (2017). Active innovation resistance: An empirical study on functional and psychological barriers to innovation adoption in different contexts. Industrial Marketing Management, 71, 95–107. https://doi.org/10.1016/j.indmarman.2017.12.011

Kafetzopoulos, D., & Skalkos, D. (2019). An audit of innovation drivers: some empirical findings in Greek agri-food firms. European Journal of Innovation Management, 22(2), 361–382. https://doi.org/10.1108/EJIM-07-2018-0155

Kaine, G. (2004). Consumer behaviour as a theory of innovation adoption in agriculture. Social research working paper, 1(4), 1-23.

Kaur, H., Künnemeyer, R., & McGlone, A. (2017). Comparison of hand-held near infrared spectrophotometers for fruit dry matter assessment. Journal of Near Infrared Spectroscopy, 25(4), 267–277. https://doi.org/10.1177/0967033517725530

Kim, D.-Y., Kumar, V., & Kumar, U. (2012). Relationship between quality management practices and innovation. Journal of Operations Management, 30(4), 295–315. https://doi.org/10.1016/j.jom.2012.02.003

Kislev, Y., & Shchori-Bachrach, N. (1973). The Process of an Innovation Cycle. American Journal of Agricultural Economics, 55(1), 28–37. https://doi.org/10.2307/1238658

Läpple, D., Renwick, A., & Thorne, F. (2015). Measuring and understanding the drivers of agricultural innovation: Evidence from Ireland. Food Policy, 51, 1–8. https://doi.org/10.1016/j.foodpol.2014.11.003

Li, M., Qian, Z., Shi, B., Medlicott, J., & East, A. (2018). Evaluating the performance of a consumer scale SCiO™ molecular sensor to predict quality of horticultural products. Postharvest Biology and Technology, 145, 183–192. https://doi.org/10.1016/j.postharvbio.2018.07.009

Magwaza, L. S., Opara, U. L., Nieuwoudt, H., Cronje, P. J. R., Saeys, W., & Nicolaï, B. (2012). NIR Spectroscopy Applications for Internal and External Quality Analysis of Citrus Fruit—A Review. Food and Bioprocess Technology, 5(2), 425–444. https://doi.org/10.1007/s11947-011-0697-1

Mani, Z., & Chouk, I. (2016). Drivers of consumers' resistance to smart products. Journal of Marketing Management, 33(1-2), 76–97. https://doi.org/10.1080/0267257X.2016.1245212

OECD (2018). OECD Fruit and Vegetables Scheme: Guidelines on objective tests to determine quality of fruit and vegetables, dry and dried produce. Available at https://www.oecd.org/agriculture/fruit-vegetables/publications/oecd-guidelines-fruit-vegetables.htm (accessed on February 27, 2020)

Onwezen, M. C., & Bartels, J. (2011). Which perceived characteristics make product innovations appealing to the consumer? A study on the acceptance of fruit innovations using cross-cultural consumer segmentation. Appetite, 57(1), 50–58. https://doi.org/10.1016/j.appet.2011.03.011

Pantano, E. (2014). Innovation drivers in retail industry. International Journal of Information Management, 34(3), 344–350. https://doi.org/10.1016/j.ijinfomgt.2014.03.002

Pantano, E., & Viassone, M. (2014). Demand pull and technology push perspective in technology-based innovations for the points of sale: The retailers evaluation. Journal of Retailing and Consumer Services, 21(1), 43–47. https://doi.org/10.1016/j.jretconser.2013.06.007

Pasquini, C. (2003). Near Infrared Spectroscopy: fundamentals, practical aspects and analytical applications. Journal of the Brazilian Chemical Society, 14(2), 198–219. https://doi.org/10.1590/S0103-50532003000200006

Pierpaoli, E., Carli, G., Pignatti, E., & Canavari, M. (2013). Drivers of Precision Agriculture Technologies Adoption: A Literature Review. Procedia Technology, 8, 61–69. https://doi.org/10.1016/j.protcy.2013.11.010

Popping, B., Allred, L., Bourdichon, F., Brunner, K., Diaz-Amigo, C., Galan-Malo, P., . . . Yeung, J. (2018). Stakeholders' Guidance Document for Consumer Analytical Devices with a Focus on Gluten and Food Allergens. Journal of AOAC International, 101(1), 185–189. https://doi.org/10.5740/jaoacint.17-0425

Popping, B., & Bourdichon, F. (2018). Food Analysis In-Depth Focus: Consumer food testing devices: threat or opportunity? Available at https://www.newfoodmagazine.com/article/64592/food-analysis-depth-focus-2018/ (accessed on February 27, 2020)

Prange, R. K., DeLong, J. M., Daniels-Lake, B. J., & Harrison, P. A. (2005). Innovation in controlled atmosphere technology. Stewart Postharvest Review, 1(3), 1–11. https://doi.org/10.2212/spr.2005.3.9

Rateni, G., Dario, P., & Cavallo, F. (2017). Smartphone-Based Food Diagnostic Technologies: A Review. Sensors (Basel, Switzerland), 17(6). https://doi.org/10.3390/s17061453

Rogers, E. M. (2003). Diffusion of Innovations (Fifth Edition). New York, NY: Free Press.

Rong, A., Akkerman, R., & Grunow, M. (2011). An optimization approach for managing fresh food quality throughout the supply chain. International Journal of Production Economics, 131(1), 421–429. https://doi.org/10.1016/j.ijpe.2009.11.026

Soderlund, R., Williams, R., & Mulligan, C. (2008). Effective adoption of agri-food assurance systems. British Food Journal, 110(8), 745–761. https://doi.org/10.1108/00070700810893296

Spectral Engines Oy (2018). FoodScanner. Available at https://www.spectralengines.com/products/nirone-scanner/foodscanner (accessed on February 27, 2020)

Sunding, D., & Zilberman, D. (2001). Chapter 4 - The agricultural innovation process: Research and technology adoption in a changing agricultural sector. Handbook of Agricultural Economics. (Volume 1, Part A), 207–261. https://doi.org/10.1016/S1574-0072(01)10007-1

Talukder, M. (2019). Causal paths to acceptance of technological innovations by individual employees. Business Process Management Journal, 25(4), 582–605. https://doi.org/10.1108/BPMJ-06-2016-0123

TellSpec Inc. (2018). Building Food Trust: Real-time, portable analysis for food safety. Available at http://tellspec.com/eng/ (accessed on February 27, 2020)

Tidd, J. (2006). A Review of Innovation Models. Imperial College London.

Trienekens, J., van Uffelen, R., Debaire, J., & Omta, O. (2008). Assessment of innovation and performance in the fruit chain. British Food Journal, 110(1), 98–127. https://doi.org/10.1108/00070700810844812

UNECE (2017). Fresh Fruit and Vegetables - Standards. Available at https://www.unece.org/trade/agr/standard/fresh/ffv-standardse.html (accessed on February 27, 2020)

Wyrwa, J., & Barska, A. (2017). Innovations in the food packaging market: active packaging. European Food Research and Technology, 243(10), 1681–1692. https://doi.org/10.1007/s00217-017-2878-2

Yu, M., & Nagurney, A. (2013). Competitive food supply chain networks with application to fresh produce. European Journal of Operational Research, 224(2), 273–282. https://doi.org/10.1016/j.ejor.2012.07.033

4 Non-destructive measurement method for a fast quality evaluation of fruit and vegetables by using food-scanner

In: DGG-Proceedings 8 (13), p. 1-5

Cite as: Goisser, S.; Fernandes, M.; Ulrichs, C.; Mempel, H. (2018): Non-destructive measurement method for a fast quality evaluation of fruit and vegetables by using food-scanner. DGG-Proceedings 8 (13), p. 1-5.

DOI: https://doi.org/10.5288/dgg-pr-sg-2018

Non-destructive measurement method for a fast quality evaluation of fruit and vegetables by using food-scanner

Simon Goisser*, Michael Fernandes, Christian Ulrichs, Heike Mempel

DGG-Proceedings, Vol. 8, 2018, No. 13, p. 1-5.
DOI: 10.5288/dgg-pr-sg-2018

*Corresponding Author:

Simon Goisser
Fakultät Gartenbau und Lebensmitteltechnologie
Hochschule Weihenstephan-Triesdorf
Am Staudengarten 10
85354 Freising
Germany

Email: simon.goisser@hswt.de

DGG-Proceedings, Vol. 8, 2018, No. 13, p. 1-5.

Non-destructive measurement method for a fast quality evaluation of fruit and vegetables by using food-scanner

Simon Goisser[1], Michael Fernandes[2], Christian Ulrichs, Heike Mempel[1]

[1]Institut für Gartenbau, Hochschule Weihenstephan-Triesdorf, Germany
[2]Technologiecampus Grafenau, Technische Hochschule Deggendorf, Germany
[3]Faculty of Life Sciences, Urban Plant Ecophysiology, Humboldt-Universität zu Berlin, Germany.

1. Introduction, Knowledge, Objectives

Recent reports estimate the volume of food loss along the supply chain to 1,3 billion tons globally per year, which equals one-third of food produced for human consumption (FAO, 2011). Further studies conducted for the german food supply chain estimate the quantity of annual food loss between 11 million (Universität Stuttgart, 2012) and 18 million (WWF, 2015) tons. Fruits and vegetables, with a percentage of 44 of the total food loss, are commodities most frequently thrown away (BMEL, 2012).
In recent years, a lot of attention is given to so-called food-scanners. Food-scanner are miniaturized near-infrared (NIR) spectrometers, which allow a fast and noninvasive determination of food quality. They can be used as a multidimensional predictor to determine the chemical and physical composition of agricultural- and food products (e.g. soluble solids, dry matter, moisture, firmness). Due to their small size and portability these devices can be used for in-field application as well as for researchers and end-consumers (Santos et al., 2013).
Studies of Flores et al. (2009) and Kim et al. (2013) indicate that NIRS is suitable for predicting quality attributes of various tomato varieties. The experiments described in this study are conducted on tomatoes and focus on the performance of a food-scanner compared to a benchtop NIR-spectrometer. Important quality parameters of tomato, such as sugar content and firmness, are evaluated with respect to their predictability in order to validate the performance of this new kind of non-destructive measurement-method.

2. Material and Methods

2.1 Materials

Salad- and cocktail-tomatoes (*Solanum lycopersicum* 'EZ 1256' and 'EZ 1359') were harvested in September and October 2017 from a greenhouse of the University of Applied Sciences Weihenstephan-Triesdorf (latitude 48°24'6"N and longitude 11°43'53"E), where they have been cultivated throughout the summer of 2017 in a run-to-waste system on rock wool. After removing the stems tomatoes were numbered with a waterproof marker at the stem basis. In a first experiment, 160 tomatoes (80 salad and 80 cocktail) were stored for three weeks to initiate post-ripening-processes and to obtain different quality levels in terms of sugar-concentration, which is expressed as total soluble solids (TSS). Ten fruits of each cultivar were then measured spectrometrically and destructively for TSS every two to three days. A second experiment was conducted with 120 tomatoes (60 salad and 60 cocktail) for the determination of firmness and dry matter (DM). Fruits were stored for two weeks to enable water loss of fruits and to obtain different levels of firmness and dry matter in tomatoes. Ten fruits of each variety were measured spectrometrically every two days followed by measurements for firmness and dry matter. Fruits in both experiments were stored at room temperature (20 °C at night - 22 °C at day) and relative humidity of 60 - 70 % under ambient light conditions at an average of 0,4 W/m² for 12h during the day.

2.2 Recording of spectra

Spectroscopic measurements were performed using a hand-held SCiO spectrometer version 1.2 (Consumer Physics, Tel Aviv, Israel) and a benchtop NIR spectrometer (Carl Zeiss MCS 621 VIS II with reflection measuring head OMK 500-H, Jena, Germany). The measurement method for both devices is diffuse reflection. Fruit spectra of tomatoes were recorded with the SCiO by taking four measurements orthogonally around the equator of each fruit (Figure 1A). For the benchtop device, spectra were acquired using a turntable (Figure 1B). Rotating fruits were scanned for six seconds. The 60 spectra recorded in this time were averaged using CORA Plus and OPC run software (Carl Zeiss, Jena, Germany).

Figure 1. Recording of spectra using the portable food-scanner SCiO (A) and a Zeiss benchtop NIR-spectrometer (B)

2.3 Acquisition of reference values

Reference measurements were made directly after recording spectra. Firmness was measured by averaging four measurements around the equator of each fruit using a non-invasive hand-held penetrometer AGROSTA 100X (Agro Technologie, Serqueux, France). Measurements were taken at the spots where spectra were acquired with the SCiO. Values are expressed in Newton by taking account of penetrometer-head-diameter and maximum force. Since fruits were not damaged and no fruit tissue was lost by this penetrometer, tomatoes were subsequently used for TSS and DM measurements.

Sugar content in terms of TSS was analyzed according to the OECD fruit and vegetables scheme by taking two longitudinal slices from opposite sides of the fruit, squeezing the slices with a garlic press and measuring the mixed juice with a digital refractometer HI 96801 (Hanna Instruments, Woonsocket, USA) in degrees Brix to one decimal place. Dry matter was measured gravimetrically for whole tomatoes. Fresh fruits were weighed and then dried in an oven at 105 °C for 48 h. The final dry weight was used to calculate DM as the percentage of dry weight to initial wet weight of each fruit.

2.4 Statistical analysis / multivariate analysis
After acquiring reference values, every averaged firmness-, TSS- and DM-value was correlated with the four SCiO spectra and the averaged Zeiss spectrum, respectively. Partial least square (PLS) models were developed with the open source statistical software R and Unscrambler (CAMO, Oslo, Norway) to establish prediction models for TSS, firmness and DM values. Sample data contains a combination of both tomato varieties (salad and cocktail) in order to generate a wider range of values. Data was preprocessed by taking the second derivative of the spectra and applying a standard normal variate correction (SNV), which led to the best result for the prediction models. The Savitzky-Golay transformation was performed with a filter width of 25. For analysis of TSS eight batches with 20 samples resulted in a total sample size of 160. Data of 120 samples, comprising six batches with 20 samples, was used for establishing models for dry matter and firmness. Cross validation for each analysis was carried out with a sample of 20 segments

3. Results

In this study, the SCiO device delivered prediction models with high linear correlations (r^2), ranging from 0.80 for dry matter, 0.82 for firmness to 0.92 for TSS. In comparison, models for the Zeiss-spectrometer showed higher values for linear correlations for every parameter examined (0.94 for TSS, 0.83 for firmness and 0,85 for dry matter). The errors of prediction, expressed as root mean square errors (RMSE), for TSS and dry matter were lower for the benchtop-device compared to the SCiO (0.37 °Brix instead of 0.45 °Brix for TSS and 0.38 % DM instead of 0.41 % DM). Firmness showed similar prediction errors (0.57 N) for both devices. Performance of models for all three traits examined in comparison to the respective spectrometer is shown in Table 1.

Table 1. Comparison of performance of SCiO handheld- and Zeiss benchtop-spectrometer for prediction of TSS, firmness and dry matter

Quality parameters	λ	n	r^2	RMSE
SCiO Handheld Device	740 - 1070 nm Resolution: 1 nm			
TSS (Brix)		160	0.917	0.453
Firmness		120	0.815	0.566
Dry Matter		120	0.802	0.412
Carl Zeiss MCS 621 VIS II	750 - 1080 nm Resolution: 2 nm			
TSS (Brix)		160	0.939	0.368
Firmness		120	0.829	0.570
Dry Matter		120	0.846	0.383

Figure 2 shows exemplarily the correlation between dry matter measured during the experiment and the obtained spectra. After calibration (red values) the model presented a

RMSE of 0.30 % DM and a linear correlation (r^2) of 0.89. Using this model and applying a second dataset (blue values) for validation, the model shows a RMSE of 0.41 % DM and a correlation of 0.80.

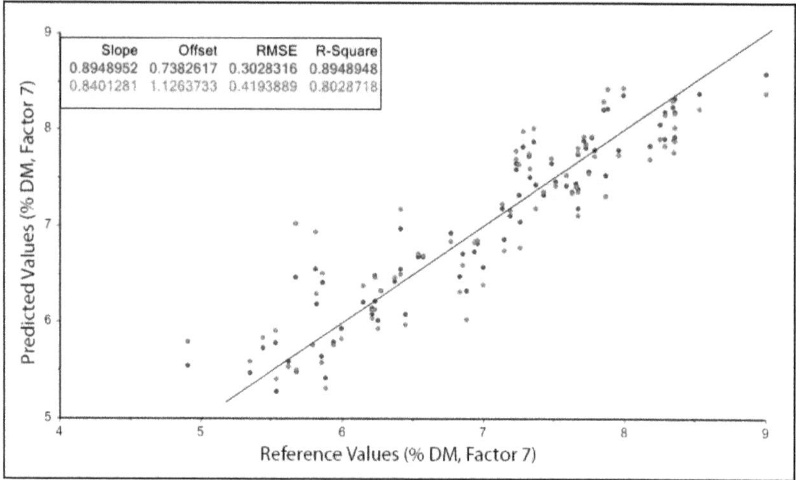

Figure 2: Accuracy of prediction of tomato dry matter using the SCiO device. Blue values were used for calibration, red values for validation of the model

4. Discussion

Main purpose of the developed PLS models was the performance-evaluation of the handheld device SCiO and the comparison to a benchtop-spectrometer. The results of both devices show that the examined quality parameters (TSS, DM, firmness) had a high correlation to collected spectra. For the SCiO device these predictions showed a slightly lower r^2 and some higher RMSE values compared to the Zeiss spectrometer. Nevertheless the obtained results indicate that the SCiO could be used for the prediction of firmness, dry matter and TSS with good to high accuracy. Similar experiments for the SCiO have only been performed with apple and kiwi (Kaur et al., 2017) and showed comparable results for the predictability of dry matter (r^2 = 0.87, RMSEP = 0.70 % DM). PLS regression performed by Kusumiyati et al. (2008) for a portable NIR spectrometer resulted in a slightly higher r^2 = 0,89 for the prediction of tomato firmness after harvest.

The models built in this study are based on two tomato varieties grown locally at the University of Applied Sciences. To be able to make accurate predictions in a more general way, e.g. by building a global model for the prediction of TSS, DM and firmness, additional tomato varieties need to be included into the existing data. Adding samples from the supermarket could also help to improve these global models by displaying quality further down the supply chain.

Additional preprocessing of data (focus on different spectral ranges, improvement of signal to noise ratio) could lead to optimization of prediction models and will be part of further studies.

5. Conclusions

First results obtained for tomatoes show promising possibilities for a prediction of fruit quality in a non-destructive way. Since more quality parameters are necessary to give a precise prediction of the degree of ripeness and maturity, additional fruit parameters (e.g. color, moisture, content of lycopene) will be included in future studies. Furthermore, other hand-held devices need to be tested to take the effect of different wavelength-ranges into account. In the future, these devices could be used to determine fruit quality on different levels of the horticultural supply chain. Supporting decisions how to proceed with different batches of fruits, food-scanner can help to reduce food loss along the supply chain

This work was supported by the Bavarian Ministry of Food, Agriculture and Forestry as part of the alliance "Wir retten Lebensmittel".

6. Literature

BMEL (2012). Initiative "Zu gut für die Tonne" - Studie über Lebensmittelabfälle in Deutschland. https://www.bmel.de/DE/Ernaehrung/ZuGutFuerDieTonne/_Texte/Studie_Zu-gut-fuer-die-Tonne.html

FAO (2011). Global food losses and food waste - Extent, causes and prevetion. Rome. http://www.fao.org/docrep/014/mb060e/mb060e00.pdf

Flores, K., Sánchez, M.-T., Pérez-Marín, D., Guerrero, J.-E. and Garrido-Varo, A. (2009). Feasibility in nirs instruments for predicting internal quality in intact tomato. *Journal of Food Engineering*, 91(2): 311-318.

Kaur, H., Künnemeyer, R. and McGlone, A. (2017). Comparison of hand-held near infrared spectrophotometers for fruit dry matter assessment. *Journal of Near Infrared Spectroscopy*, 25(4): 267-277.

Kim, G., Kim, D.-Y., Kim, G. H., and Cho, B.-K. (2013). Applications of discrete wavelet analysis for predicting internal quality of cherry tomatoes using vis/nir spectroscopy. *Journal of Biosystems Engineering*, 38(1): 48-54.

Kusumiyati, Akinaga, T., Tanaka, M., and Kawasaki, S. (2008). On-tree and after-harvesting evaluation of firmness, color and lycopene content of tomato fruit using portable NIR spectroscopy. *Journal of Food, Agriculture & Environment*, 6(2): 327-332.

Santos, C. A., Lopo, M., Páscoa, R. N., & Lopes, J. A. (2013). A Review on the Applications of Portable Near-Infrared Spectrometers in the Agro-Food Industry. *Applied Spectroscopy*, 1215-1233.

Universität Stuttgart (2012). Ermittlung der weggeworfenen Lebensmittelmengen und Vorschläge zur Verminderung der Wegwerfrate bei Lebensmitteln in Deutschland

WWF (2015). Geschätzte Lebensmittelverluste in Deutschland nach Wertschöpfungsstufen im Jahr 2014 (in Mio. Tonnen). In *Statista - Das Statistik-Portal*

5 Evaluating the practicability of commercial food-scanners for non-destructive quality assessment of tomato fruit

In: Journal of Applied Botany and Food Quality 93, p. 204-214

Cite as: Goisser, S.; Fernandes, M.; Wittmann, S.; Ulrichs, C.; Mempel, H. (2020): Evaluating the practicability of commercial food-scanners for non-destructive quality assessment of tomato fruit. Journal of Applied Botany and Food Quality 93, p. 204–214.

DOI: https://doi.org/10.5073/JABFQ.2020.093.025

Title

Evaluating the practicability of commercial food-scanners for non-destructive quality assessment of tomato fruit

Short title

Non-destructive quality assessment using food-scanners

Author names and affiliations

Simon Goisser[1,3], Michael Fernandes[2], Sabine Wittmann[1], Christian Ulrichs[3], Heike Mempel[1]*

[1]Greenhouse Technology and Quality Management, University of Applied Sciences Weihenstephan-Triesdorf, Freising, Germany

[2]Department of Applied Artificial Intelligence, Deggendorf Institute of Technology, Technology Campus Grafenau, Grafenau, Germany

[3]Division Urban Plant Ecophysiology, Faculty of Life Sciences, Humboldt-Universität zu Berlin, Berlin, Germany

* Corresponding author

Summary

The assessment of tomato fruit quality depends on a variety of extrinsic and intrinsic quality parameters such as color, firmness and sugar content. Conventional measurement methods of these quality parameters are time consuming, require various measurement devices, and in case of intrinsic quality, involve destructive measurements. Latest research focused on the non-destructive determination of these parameters by using spectroscopic measurements. The goal of this study was to evaluate the capability of three commercially available portable and miniaturized Vis/NIR spectrometers, so called food-scanners, in predicting various tomato quality attributes in a non-destructive way. Additionally, this study evaluated the software provided by manufacturers for building of prediction models by comparing the results derived from those software tools to state-of-the-art software for multivariate analysis. Evaluation of food-scanner spectra resulted in prediction models of high accuracy ($r^2 > 0.90$) for tomato fruit firmness, dry matter, total soluble solids and color values L*, a* and h°. Prediction models computed with manufacturer's software showed similar accuracy to those derived from state-of-the art evaluation software. Results of this study illustrate the great potential of commercial food-scanners for non-destructive quality measurement. Further important features of food-scanners with respect to the application along the fresh produce supply chain are addressed.

Keywords

NIR spectroscopy; tomato; quality control; non-destructive measurement; food-scanner

Introduction

Fresh tomato fruit (*Solanum lycopersicum* L.) is the vegetable species most consumed in Germany in terms of quantity with an average amount of 2.3 Mio t per year (STATISTA, 2019). When it comes to cultivation, distribution and marketing of tomatoes, various commercial quality parameters are of high relevance. These commercial quality attributes differ depending on the position in the supply chain. Producer prioritize qualities beneficial for an unproblematic cultivation of plants such as disease resistance, cultivation work, fruit weight, fruit size and appearance. In order to meet the requirements of a demand driven supply chain, additional postharvest characteristics such as uniformity, shape, color and firmness as well as shelf life are of importance (FOLTA and KLEE, 2016). From the perspective of the consumer, taste and flavor are relevant quality attributes, and lack of flavor was identified as the primary reason for consumer dissatisfaction for tomatoes (BRUHN et al., 1991). A recent study on consumer acceptance of tomatoes cultivated for fresh consumption attested the importance of sensory traits like juiciness, sweetness and taste intensity, which influence consumer purchase preference (CASALS et al., 2019). Additionally, tomato firmness and color are vital quality parameters which influence acceptability and marketability (BATU, 2004).

Some of the above mentioned quality parameters of tomato fruit must meet certain standards within the European Union regarding the marketing and commercial quality control (BLE, 2019). Besides these standards required by law, retail companies impose additional requirements which oftentimes exceed those statutory provisions. These provisions regarding quality relate primarily to extrinsic attributes that are important for marketing and trading, e.g. appearance or absence of damage and deterioration. Firmness is used as an indicator for the classification of tomatoes into three classes (Extra, Class I, Class II), yet the quoted indications "firm, reasonably firm, slightly less firm than Class I" and severity of defects (UNECE, 2018) are prone to subjectivity. Visual and haptical impressions dominate quality assessment along the fresh fruit supply chain, therefore the evaluation of these traits highly depends on experience of staff in quality control (GOISSER et al., 2020a). The application of instrumental measurements for these commercially important traits could help to reduce the variation between individual controller, offer higher precision and supply a standardized language of fruit quality among industry, consumers and research (ABBOTT, 1999). Furthermore, the measurement of intrinsic quality parameters such as sugar content (°Brix) as well as acidity, which relate to tomato flavor and sweetness (CAUSSE et al., 2007), could allow the estimation of sensory traits such as tomato taste.

Traditional instrumental measurement methods of extrinsic and intrinsic quality parameters are often time consuming and require trained personal in combination with various measurement devices. The determination of firmness via penetrometer and sugar content per refractometer demands destructive measurement methods, in the case of acidity handling of chemicals is necessary for titration. Besides these destructive measurement methods of internal quality parameters, optical measurement methods like visible and near infrared (VIS/NIR) spectroscopy have been successful in determining important quality parameters of agricultural and horticultural products. NIR spectroscopy can be applied for qualitative applications, such as identification and classification of samples, as well as quantitative applications in order to determine major constituents in samples such as fruit and vegetables (PASQUINI, 2003). As emphasized in the recent guidelines on objective tests for fruit and vegetables (OECD, 2018), NIR spectroscopy comprises the ability to simultaneously predict multiple quality attributes using only a single spectrum, therefore serving as multidimensional predictor of fruit quality. A summary provided by DOS SANTOS et al. (2013) highlights successful reported applications for analysis of various quality attributes along a variety of fruit and vegetable species using portable spectrometers instead of traditional laboratory NIR devices. Ongoing technological developments promoted the miniaturization and commercialization of NIR sensors, so called food-scanners. Some of these devices are particularly designed for end-consumers and should enable them to identify macronutrients, allergens, calories or food contaminants (RATENI et al., 2017).

First studies on the predictability of individual tomato quality parameters such as sugar content and lycopene (SHENG et al., 2019; GOISSER et al., 2020b) using portable handheld NIR spectrometers indicate the potential of these miniaturized devices for non-destructive quality evaluation. Commercially manufactured portable NIR instruments are relatively new and previous research either compared the performance of various instruments for the prediction of single quality traits such as dry matter (KAUR et al., 2017) or used individual devices for the prediction of multiple quality attributes within selected fruit such as kiwi (LI et al., 2018) or avocado (NCAMA et al., 2018). The results of these studies illustrate the potential of food-scanners for fast and non-destructive quality measurement. With regard to the practical use of these devices along the fresh produce supply chain, a recent study identified important requirements as stated by supply chain actors, e.g., ease of use and robustness of devices and reliability and accuracy of prediction models (GOISSER et al., 2020a).

The aim of the study is to examine the assessment of the most important quality attributes of tomato fruit by using three commercially available portable food-scanners. Contrary to previous research, which focused on prediction accuracy and spectral pre-processing, this study investigates user friendliness and potential of an application in practice along the fresh fruit supply chain. Therefore, this study used manufacturer's software for building of prediction models as well as state-of-the-art software for multivariate analysis. By comparing the results derived from both software tools, the accuracy and reliability of prediction models computed with manufacturer's software can be assessed. For the evaluation of the full potential of each food-scanner, the full available spectral range was used during building of NIR prediction models. In order to generate a high variability for different quality parameters, two different experimental approaches were used in this study. In one experiment, different ripening stages from one tomato cultivar were examined, whereas the second experiment analyzed tomatoes of uniform ripening levels but different cultivars and origins. These different approaches highlight the fact that users have to adhere to certain requirements in order to create good prediction models. Further important aspects with respect to usability and applicability by non-experts are discussed.

Materials and Methods

Experimental setup and sample material

All research was conducted at facilities of the University of Applied Sciences Weihenstephan-Triesdorf. In order to examine tomato fruit quality in its entirety the experiment comprised two separate parts. Within the first part, quality of one tomato variety (*Solanum lycopersicum* cv. Avalantino) was monitored from fruit ripening during cultivation throughout harvest up to postharvest storage. Tomatoes were cultivated during summer 2019 in a greenhouse at the University of Applied Sciences Weihenstephan-Triesdorf (48°24'12"N, 11°43'50"E) under commercial growth conditions. A hydroponic recirculating drip irrigation system was utilized for tomato cultivation. The average air temperature during cultivation within the greenhouse was 20.1 ± 8.1 °C and relative humidity was 74.5 ± 17.7 %. In total, 245 tomatoes were used for the first part of the experiment. Evaluation of tomato quality during ripening was conducted from the middle of July until start of August 2019 on five days with two to five days in between. Five different ripening stages were determined for harvest by using the USDA Color Chart (USDA, 1975) as orientation: 1) Green (tomato surface of full green color) 2) Breaker (shift in color from green to yellow) 3) Turning (break in color from yellow to orange) 4) Light red (change in color from orange to light red) 5) Red (tomato surface of full red color). On any measurement day five tomatoes of every ripening stage were randomly selected and harvested, resulting in batches of 25 tomatoes each day and 125 fruit total. Calyx was removed and samples were numbered for recording of spectra and subsequent reference measurements. In order to monitor fruit quality during postharvest storage, 120 tomatoes of the ripening stage (5) were randomly selected and harvested from the greenhouse in the fourth week of August, 2019. After numbering the tomatoes were positioned on a metal grid to allow full air circulation at room temperature of 22.4 ± 1.7 °C and relative humidity of 67.4 ± 3.0 % for the storage period of 22 days. Measurement of fruit quality was conducted on the day of harvest and 8, 13, 17, 20, and 22 days after harvest in batches of 20 randomly selected samples.

For the second part, 200 tomatoes consisting of eight different varieties à 25 fruit were purchased at four different supermarkets in Freising, Germany, from January to March, 2019. Only plastic-wrapped tomatoes of commercial grade one were selected. In order to generate variability, various tomato types (Roma, Salad, Cherry) from different countries of origin were analyzed (Table 1). Not all fruit cultivars could be identified due to missing indications on packaging labels. All spectral and related reference measurements in both parts of the experiment were conducted on a single fruit basis. Prior to spectroscopic and reference analysis fruit was kept at room temperature to allow acclimatization. Temperature of fruit surface was measured immediately before recording of spectra using an infrared meter (AMiR 7834, Ahlborn Meßtechnik GmbH, Holzkirchen, Germany) and averaged 19.1 ± 0.4 °C.

Acquisition of spectra using portable NIR devices

Fruit spectra of intact tomato fruit were recorded with three different portable and commercially available food-scanners (Figure 1). At first, two spectra for each tomato were acquired on opposite sides of the fruit equator using the F-750 Produce Quality Meter (Firmware v.1.2.0 build 7041, Felix Instruments, Portland, USA). During each recording of a new measurement, the F-750 uses a reference shutter for normalization of the spectrometer output and accounts for dark current and ambient light by recording dark scans. After recording of spectra data was transferred from the SD-Card to a computer for subsequent analysis. In a next step, the portable NIR spectrometer H-100F (Sunforest, Incheon, Korea) was used for recording of spectra on same two spots on opposite sides of the fruit equator as the F-750. To ensure scan quality, the white standard attached in the protective cap of the measurement head was used for referencing at the beginning and after every 20 measurements. Raw spectra was saved to a computer by linking the H-100F via USB and running the Sunforest H-100 Laboratory Program (V 1.1, Sunforest, Incheon, Korea). At last, spectra of each tomato was recorded using the SCiO™ Molecular Sensor (SCiO™ version 1.2, Consumer Physics, Hod HaSharon, Israel). The SCiO™ applies a LED light source which illuminates a considerably smaller surface area compared to the halogen lamps of the F-750 and H-100F. In order to depict an appropriate average fruit spectra, four equidistant scan points (approximately 90° apart) around the fruit equator were deemed more suitable, covering the two measurement points used with the F-750 and H-100F. The device possesses a white standard within its protective plastic case, which was used for referencing at the start and after every 20 measurements. The technical features of all three devices have been elaborated in several previous works (KAUR et al., 2017; LI et al., 2018).

Depending on the respective device, two to four spectra were recorded for each fruit. For each device, all spectra belonging to one tomato fruit were averaged and the averaged spectra used for correlation with quality parameters and subsequent model building.

Fig. 1: Commercially available portable NIR food-scanner: (left) SCiO™ Molecular Sensor with protective plastic case, (middle) F-750 Produce Quality Meter, (right) H-100F portable spectrometer with protective plastic cap

Measurement of fruit quality parameters

Measurements of external and internal quality parameters were performed immediately after recording of spectra and followed a defined operating procedure (Figure 2). At first, fruit color was measured around the fruit equator with a colorimeter (PCE-CSM 2, PCE Instruments, Meschede, Germany) consistent with the spots of NIR spectra acquisition. Color readings were automatically averaged by the colorimeter and the averaged values used for subsequent reference. Firmness of each fruit was recorded in N as an average of two measuring points using a penetrometer (Fruit Texture Analyser, GUESS Manufacturing Ltd., Cape Town, South

Africa). The applied probe head was 1 cm² in surface and used a measurement speed of 5 mm/s and test depth of 7.5 mm.

After acquisition of external quality parameters fruit were randomly selected for dry matter (DM) measurement. Within the first part of the experiment, dry matter of 50 tomatoes was analyzed during the ripening period, consisting of two tomatoes of each of the five ripening stages on five measurement days. During analysis of stored fruit, five tomatoes were selected randomly for DM measurement on each of the six days of measurement, resulting in 30 tomatoes. In total, 80 samples were analyzed for DM within the first part. In the second part, ten tomatoes within the batch of 25 fruit of each of the eight varieties were drafted randomly for DM measurement, resulting in 80 samples in total. Tomatoes were cut into four parts, put in small aluminum containers and placed in a dry oven for 48 h at 80 °C until constant dry weight was reached. Afterwards the ratio of dry weight to initial fresh weight was used to calculate % DM.

The remaining tomatoes within each part of the experiment were blended using a conventional smoothie mixer (WMF Kult X Mix&Go 300 Watt, WMF Group GmbH, Geislingen/Steige, Germany). The purée was filtered using filtering paper. 2-3 drops of stirred filtrate were used for measurement of total soluble solids (TSS) using a digital refractometer (HI 96801, Hanna Instruments, Woonsocket, USA). 10 mL stirred filtrate was used for the determination of titratable acidity with a titrator (HI 902 Potentiometric Titrator and HI 921 Autosampler, Hanna Instruments, Woonsocket, USA). Results were expressed as g/L citric acid in fresh weight. TSS concentration and amount of titratable acidity were used to calculate brix/acid ratio as stipulated by OECD guidelines (OECD, 2018).

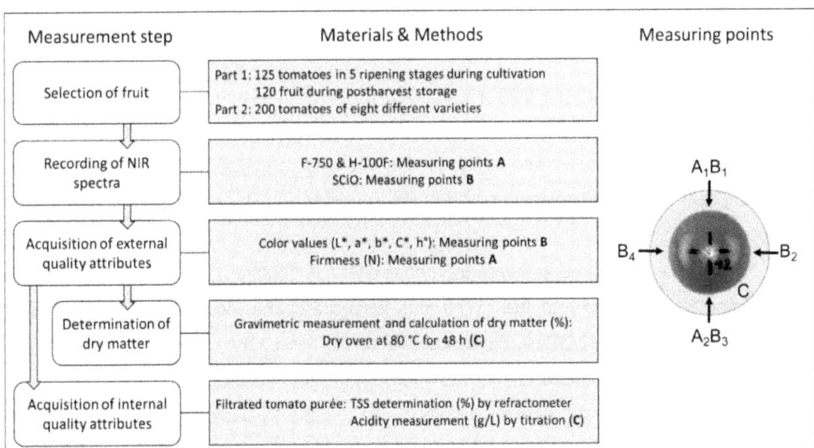

Fig. 2: Illustration of the measurement procedure and measuring points for NIR spectra as well as quality parameters within this experiment

Statistical and chemometrical analysis

General data analysis such as calculation of average values and standard deviations (SD) was performed with Microsoft® Excel® 2016.

The evaluation of a linear relationship between recorded spectra and tomato quality attributes was performed with the respective analysis software tools from the device manufacturer for the SCiO™ and F-750. In order to evaluate the performance of the manufacturer's software and to carry out a neutral overall evaluation of food-scanner predictability of quality parameters, raw spectra of both devices was additionally analyzed with the multivariate data analysis software The Unscrambler® software (version 10.5.1, CAMO, Oslo, Norway). In comparison, the H-100F is not provided with an end-user tool for creating prediction models, therefore spectra was only analyzed with The Unscrambler® software. Software such as The Unscrambler® allow the computation of the correlation coefficient of both calibration (r^2_C) and cross-validation model (r^2_{CV}) as well as the corresponding calibration and cross-validation root mean square error ($RMSE_C$, $RMSE_{CV}$). These factors are used to evaluate the performance of the respective prediction model. Additionally, the software offers numerous possibilities for data pre-processing and further analysis. Outliers detected with The Unscrambler®, which indicated errors during spectra collection, were omitted from evaluation.

After recording of spectra with the SCiO™, the spectra are uploaded to a cloud-based browser application (The Lab, Consumer Physics, Hod HaSharon, Israel) which allows building of prediction models. The default pre-processing settings within the cloud-application were used in order to build prediction models, namely logarithm, averaging all four scans per sample, first derivative (window of 35 and polynom degree of 2), selecting the full wavelength range from 740 - 1070 nm and subtracting the average. Every tenth spectra was used for cross-validation. The algorithm was set to partial least square regression (PLSR) and outlier detection as well as additional filters were disabled. Raw spectra were then exported from the cloud-based samples library to an Excel-file and analyzed with The Unscrambler®, applying the same pre-processing settings as set in the cloud-application as well as PLSR, using 20 random segments for cross-validation.

The F-750 comes with a free software tool for analysis of spectra and building of prediction models. Spectra recorded with the F-750 was loaded into the most recent Model Builder Software (version 1.3.0.192 BETA, Felix Instruments, Portland, USA). The software uses non-linear iterative partial least squares (NIPALS) regression and leave-one-out method for cross-validation. Spectra in second derivative form, which is used by default by the software, was transferred from the Model Builder Software to an Excel-file and loaded into The Unscrambler® for additional evaluation. The whole wavelength range from 477 – 1059 nm was utilized for building prediction models. Both spectra per sample were averaged and no further pre-processing was applied. PLSR algorithm was utilized for building of prediction models, using

20 random segments for cross-validation.

For the H-100F, raw spectra in second derivative form was exported from The Sunforest H-100 Laboratory Program in the wavelength range from 650 - 950 nm. After averaging the two scans per sample no further pre-processing was applied. 20 random segments were used for cross-validation.

The SciO™ application as well as the F-750 Model Builder Software allow the selection of a specific spectral range for evaluation. Within this study, the full wavelength range of all food-scanners was selected in order to evaluate the full potential of these devices. Since both software programs of the F-750 and H-100F do not allow access to raw spectra, the automatically computed second derivative of spectra was used.

Gathering of information on usability characteristics

A qualitative study regarding the application of food-scanners along the fresh produce supply chain identified several important requirements for food-scanners to allow the utilization in daily quality control processes, e.g., high prediction accuracy, convenience in handling and robustness of food-scanners (GOISSER et al., 2020a). Based on these results as well as personal experience of the authors through collaborations with supply chain actors, characteristics of high importance with respect to the usability and applicability of food-scanners were identified within four categories, namely device independency, data handling, practical handling and available accessories. Manufacturer's specifications were used to evaluate device independency and available accessories, whereas personal assessments of the authors were used besides manufacturer's specifications for the evaluation of data handling and practical handling of food-scanners. In order to determine a practice-oriented scan speed for each device, 30 consecutive measurements on various tomatoes were conducted with each food-scanner and measured using a stop watch. Based on these measurements, arithmetic mean as well as standard deviation of scan speed for each device were calculated.

Results

Distribution of tomato fruit quality for both parts of the experiment

Value ranges, arithmetic mean and standard deviation (SD) for all quality parameters measured in both parts of the experiment are shown in Table 2. As indicated by the value range, variance within the first part of the experiment was greater for quality parameters L*, a*, C*, h°, firmness and brix/acid ratio. Vice versa, quality attributes such as DM, TSS and titratable acidity showed higher variability within the second part of the experiment. Variance of color value b* differed only slightly between both parts of the experiment.

Food-scanner spectra vs. quality parameters (manufacturer's software)

The Model Builder software allocated by Felix Instruments for the F-750 provides the user with information on model linearity of the calibration set (r^2_C) as well as the correlation coefficient after performing a leave-on-out cross-validation (r^2_{CV}). Root mean square errors for both calculations are also provided ($RMSE_C$, $RMSE_{CV}$). For the purpose of this study, these values were chosen as denominator of performance (Table 3). The cloud application provided by Consumer Physics for the evaluation of SCiO™ spectra with respect to its correlation to reference values does not differentiate between calibration and validation models and therefore only displays a single r^2 and RMSE value (Table 3).
Evaluation of spectra collected with the F-750 in the first part of the experiment yielded models of high prediction accuracy ($r^2_{CV} > 0.90$) for color values L*, a*, h° as well as firmness. Prediction models of moderate performance were obtained for chroma C* ($r^2_{CV} = 0.72$) as well as TSS ($r^2_{CV} = 0.56$), whereas color value b* and Brix/Acid ratio yielded predictions of poor performance. No feasible calibration and cross-validation could be acquired for DM ($r^2_{CV} = 0.03$) and titratable acidity ($r^2_{CV} = 0.02$). Within the second part of the experiment, prediction models computed for DM and TSS showed high predictive capabilities ($r^2_{CV} > 0.90$). Models of moderate performance were acquired for color value L* ($r^2_{CV} = 0.52$) and b* ($r^2_{CV} = 0.65$), whereas predictions of color value a*, chroma C*, hue h°, firmness, titratable acidity and Brix/Acid ratio displayed poor accuracy ($r^2_{CV} < 0.50$).
Prediction models of high accuracy ($r^2_P > 0.90$) computed for the SCiO™ device in the first experiment part were obtained for color value a*, hue h° and firmness, while models for L*, chroma C* and TSS were of good performance with r^2_P of 0.89, 0.80 and 0.71, respectively. Models of poor predictive capability were acquired color value b* ($r^2_P = 0.42$) and Brix/Acid ratio ($r^2_P = 0.49$), whereas the prediction of DM and titratable acidity was not feasible. Analysis of spectra gathered in the second part of the experiment resulted in prediction models of high accuracy ($r^2_P = 0.97$) for DM and TSS. Moderate r^2_P were obtained for color value b*, chroma C*, titratable acidity and Brix/Acid ratio, however the prediction models computed for color values L*, a*, hue h° as well as firmness showed low performance.

Food-scanner spectra vs. quality parameters (The Unscrambler®)

The evaluation of raw spectra of all three commercially available food-scanners using The Unscrambler® software showed great differences in the accuracy of quantitative prediction of certain tomato quality attributes (Table 4).

Within the first part of the experiment, tomato fruit quality was monitored from fruit ripening during cultivation up to postharvest storage. The evaluation of all three portable food-scanners within this part of the experiment resulted in cross-validation models of high accuracy (r^2_{CV} >0.90) for color values L* and a*, hue h° as well as firmness (Figure 3A). Validated models of color value b* showed low to moderate accuracy for the F-750 (r^2_{CV} = 0.29), SCiO™ (r^2_{CV} = 0.55) and H-100F (r^2_{CV} = 0.48) instrument. Prediction models of chroma C* yielded good results for all three devices, ranging from r^2_{CV} = 0.73 to 0.80 and 0.83 for F-750, H-100F and SCiO™ respectively. Unlike the second part of the experiment, calibration models for dry matter indicated only small predictive capability for all three instruments, and subsequent cross-validation resulted in a significant drop in r^2, which overall led to prediction models of poor performance. The validation of models for TSS achieved predictions of moderate (r F-750 and H-100F) to good (SCiO™) performance, whereas cross-validation of prediction models for titratable acidity only yielded models of moderate accuracy for all three food-scanners. Prediction models for the Brix/Acid ratio ranged from r^2_{CV} = 0.69 over 0.70 to 0.74 for the F-750, H-100F and SCiO™, respectively, indicating moderate to good predictability.

In the second part of the experiment, fruit quality of eight different tomato varieties was determined. The evaluation of prediction models computed with food-scanner scans and reference measurements yielded models of high accuracy (r^2_{CV} > 0.90) for both dry matter and TSS (Figure 3). Additionally, all three sensors obtained cross-validated prediction models of moderate accuracy (r^2_{CV} ranging from 0.54 to 0.75) for color values L* and b*, hue h°, firmness, acidity as well as the Brix/Acid ratio for all three devices. Low to moderate predictability was achieved for color value a*, where r^2_{CV} ranged from 0.48 over 0.49 to 0.54 for F-750, SCiO™ and H-100F, respectively. Prediction of chroma C* was poor for the F-750 (r^2_{CV} = 0.42) as well as the H-100F (r^2_{CV} = 0.49), whereas the evaluation of the SCiO™ device yielded moderate results (r^2_{CV} = 0.65).

Differences in usability characteristics

Important features with respect to device independency, data handling, practical handling and available accessories of food-scanners derived from manufacturer's specifications and personal assessment of the authors are compiled in Table 5.

With respect to the independent use of these devices without additional equipment, the necessity of internet access and additional devices during scanning as well as the need for external white-referencing of these spectrometers was evaluated. The investigation showed that the F-750 and H-100F can be used as stand-alone and offline instruments that require no additional devices for scanning or displaying of results. Both instruments are equipped with

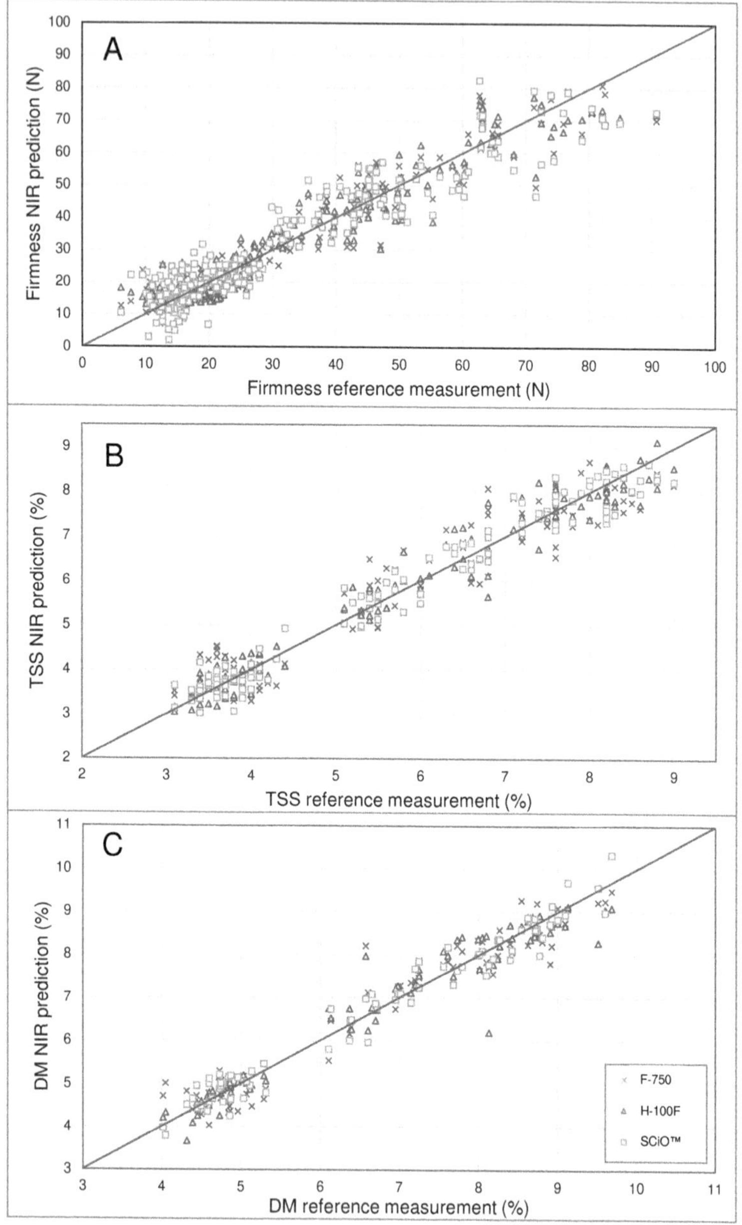

Fig. 3: Correlation between measured and predicted firmness (A), TSS (B), and DM (C) of cross validated models built with The Unscrambler® of all three food-scanners used in this study. The black line symbolizes the line of best fit.

on-board prediction models for evaluation of NIR spectra and displays for the illustration of the respective measurement results. In contrast, the SCiO™ requires constant internet access during scanning as well as an additional mobile device. For the SCiO™, an additional device (e.g., smartphone, tablet) is mandatory, works as an intermediate link between the miniaturized NIR spectrometer and the cloud database, and serves as display for measurement results. The SCiO™ as well as the H-100F provide protective plastic caps, which on the one hand serve as protection during transportation and on the other hand contain the respective white standard for regular referencing of the spectrometer. Contrary to this, the F-750 uses a shutter with gold-coated foil as its reference standard, which is calculated for every sample measurement, therefore an external white reference is not needed.

As a second aspect, the handling of the generated data was examined. Spectra collected with the F-750 can be accessed either by removing the SD card on which spectra is stored and transferring the data to a computer or using the Wi-Fi function of the SD card. Although a special file format is used for the spectra, the transfer to other multivariate analysis software is easily possible due to the free software and open format. SCiO™ spectra is only accessible through an online database by using a web browser or the corresponding SCiO™ mobile app. Furthermore, the SCiO™ requires a separate development license for the access of raw spectra, which can be downloaded in a simple file format. Spectral data stored on the H-100F can either be transferred via USB to a computer or accessed via Bluetooth using a smartphone. Spectra provided by the H-100F can either be accessed with special software for multivariate data analysis, e.g., The Unscrambler®, or by using data analysis software such as R or Python combined with syntax analysis of the respective file format. The ability of building own prediction models and the knowledge necessary for doing so differs significantly between the three devices. The manufacturer of the F-750 provides a free software tool for analysis of spectra and building of own prediction models and supports users with a systematic instruction on how to build custom tailored models. By following these instructions step by step, unexperienced users are able to build own prediction models without specialized knowledge in NIR spectroscopy. As for the SCiO™, the manufacturer provides a cloud-based browser application that allows the building and evaluation of prediction models. However, the utilization of these self-developed prediction models for future predictions requires the programming of separate mobile applications. In contrast to this, building of own prediction models by end-users is not possible with the H-100F.

Additional differences were identified with respect to features in practical handling of the three food-scanners. Regular handling of fruit and vegetables leads to light soiling of food-scanners, e.g., through adhering water and dirt or particles of rough and hairy fruit skins. Therefore, lens and light source of the SCiO™ and H-100F device have to be cleaned regularly to guarantee undistorted measurements. Due to the design of the F-750, the lens protects the light source, therefore only the lens requires regular cleaning. According to manufacturer's specifications, the F-750 battery lasts over 1600 scans and the H-100F battery over 2000 scans. Results of our own measurements estimate the battery life of the SCiO™ at 250 scans. Own measurements identified differences in scan speed, ranging from 4.9 s (H-100F), 8.2 s (SCiO™) to 14.3 s (F-750). It should be noted that the whole time interval from the beginning of one to the beginning of the next possible scan was recorded instead of just measuring the time until the measured value was displayed. Due to the built-in technology and the focus on different end-user groups, there are also differences in the weight and dimensions of the three food-scanners. The F-750 is robustly designed for the utilization within the orchard through farmers and producers, whereas the SCiO™'s small and lightweight design was initially targeted for end-consumer use. The H-100F has similar dimensions to the F-750, but since plastic is used for the casing instead of metal, it weighs significantly less. All three devices provide accessories for the measurement of smaller objects and fruit. In addition, the H-100F features a built-in radio-frequency identification (RFID) reader for the identification of individual tagged trees or areas, whereas the F-750 has a GPS sensor for online mapping of non-destructive quality readings. Manufacturers offer supplementary accessories for additional areas of application such as liquids (F-750, SCiO™) and powders (F-750).

Discussion

Prediction results of food-scanner spectra

By using two different experimental designs (tomatoes during ripening and different tomato cultivars from the supermarket) we were able to generate great variance for almost every quality parameter examined. Color values L*, a*, C*, h° and firmness showed great variation in reference values in the first part of the experiment due to ripening-related changes during cultivation and storage. The combination of different tomato cultivars in the second part of the experiment resulted in a wide range of reference values for DM and TSS. The evaluation of NIR prediction models using food-scanners yielded a higher coefficient of determination (r^2) for the respective part of the experiment with a wide range in fruit quality of the samples used in calibration and validation sets (Table 2). These findings are in line with SU et al. (2014) and highlight the importance of the variability of samples for building a feasible NIR prediction model of the quality trait of interest. This wide range of reference values also makes prediction models more robust, since both bad and good qualities are integrated. Thus a realistic range of values is represented. When creating future prediction models during practical application of food-scanners along the fresh fruit supply chain, attention should be paid to a large variance of the desired quality trait of interest. In order to guarantee this variance, users responsible for building prediction models must have extensive knowledge of fruit quality.

For the following comparison of our results to scientific literature, results computed with the independent software The Unscrambler® are used as reference. The models with the best accuracy due to the wide range of reference values are used for comparison (Table 3). Utilizing NIR spectra for the determination of tomato firmness during cultivation and storage resulted in good prediction models ($r^2_{CV} > 0.89$) for all three food-scanners. These results are superior to findings achieved on intact tomatoes using portable and laboratory NIR instruments reaching from $r^2_{CV} = 0.78$ (ECARNOT et al., 2013) to $r^2_{CV} = 0.82$ (HE et al., 2005). Similar results for firmness prediction ($r^2 = 0.89$) on intact tomatoes were achieved by KUSUMIYATI et al. (2008) using portable NIR spectroscopy.

Tomato dry matter content varies greatly depending on the tomato cultivar (ERCOLANO et al., 2008). Screening of one tomato cultivar from cultivation to storage in our study showed a smaller range of dry matter distribution (2.08 %) compared to screening multiple tomato cultivars (5.67 %). Prediction of dry matter with only one tomato cultivar yielded no feasible models ($r^2_{CV} = 0.16 - 0.32$). These findings are in line with poor prediction results ($r^2_{CV} = 0.39 - 0.49$) for one tomato cultivar using a laboratory NIR instrument (TORRES et al., 2015). Contrary to this, NIR prediction models of high accuracy ($r^2_{CV} > 0.90$) could be created for the model containing multiple cultivars with all three food-scanners. Prediction of dry matter in apple, kiwifruit and stone fruit applying the same food-scanners we used showed similar predictive capabilities of this fruit trait (KAUR et al., 2017).

Models containing only one cultivar yielded moderate predictions of TSS ($r^2_{CV} = 0.67 - 0.81$), which is in range of previous studies using only one tomato cultivar and laboratory NIR instruments (FLORES et al., 2009; TORRES et al., 2015; CLÉMENT et al., 2008). Building of TSS prediction models using multiple tomato cultivars in our study yielded accurate results ($r^2_{CV} > 0.90$). Previous studies using laboratory NIR instruments (HE et al., 2005; SAAD et al., 2016) and a miniaturized NIR sensor (SHENG et al., 2019) obtained similar prediction models.

In the first part of the experiment all three food-scanners yielded prediction models of high accuracy for lightness L* and color value a* ($r^2_{CV} > 0.90$, Table 3), which is in the range of findings from previous work using benchtop UV-VIS-NIR spectrometer and portable NIR instrument (CLÉMENT et al., 2008; KUSUMIYATI et al., 2008). Prediction of color value b* yielded best results in the second part of the experiment for all three food-scanners ($r^2_{CV} = 0.66 - 0.73$). However, previous studies demonstrated better correlations of $r^2 = 0.82 - 0.92$ (KUSUMIYATI et al., 2008; CLÉMENT et al., 2008). Correlation of food-scanner spectra to chroma C* in our study obtained moderate to good results ($r^2_{CV} = 0.73 - 0.83$) in the first part of the experiment. Various previous research using portable NIR spectrometer yielded slightly better prediction models of $r^2 = 0.90 - 0.87$ (ECARNOT et al., 2013; KUSUMIYATI et al., 2008). Evaluation of hue h° yielded prediction models of high accuracy ($r^2_{CV} > 0.90$), which is similar to findings in the literature on portable NIR instrument performance (ECARNOT et al., 2013; KUSUMIYATI et al., 2008).

Previous studies applying laboratory NIR instruments demonstrated the difficulty of predicting acidity in intact tomato fruit (FLORES et al., 2009; OLIVEIRA et al., 2014) and yielded prediction accuracies comparable to our results. As elaborated by OLIVEIRA et al. (2014), tomatoes generally show low concentration and a heterogeneous composition of TSS and acidity. Compared to other fruit species, tomatoes have no homogeneous composition of flesh since they are divided into different loculi. This structure can cause interference when NIR radiation is penetrating fruit and could be the reason for these moderate correlations in both parts of the experiment (r^2_{CV} = 0.51 - 0.75).

This moderate predictability of acidity combined with the mathematical relation yielded prediction models for brix/acid ratio of likewise moderate accuracy (r^2_{CV} = 0.54 - 0.74) in both experiments. No literature reports could be found regarding prediction models for brix/acid ratio in intact tomatoes. However, previous research on tomato juice using a laboratory NIR instrument obtained slightly superior accuracies (r^2 = 0.74 - 0.86) for comparable wavelength regions (JHA and MATSUOKA, 2004).

All three food-scanners had similar accuracies for the prediction of each quality trait within each part of the experiment (Table 3). The slight differences in model accuracy can be explained by the differences in wavelength ranges of the respective scanners. KAUR et al. (2017) evaluated the spectral features of all three food-scanners used in this study and found strong correspondence to the water absorbance bands at 760 nm, 840 nm and 960 nm, with slight variations between instruments. As demonstrated by MCGLONE and KAWANO (1998) using kiwifruit, the water and carbohydrate absorbance bands at 840 nm and 960 nm strongly correlate to prediction models developed for TSS and DM. The good correlation of food-scanner spectra and tomato TSS and dry matter in our study are in line with these findings. With respect to the determination of color values, the F-750 covers almost the whole visible spectrum up to near infrared (477 - 1059 nm), whereas the H-100F comprises the red wavelength range of the visible spectrum (650 - 950 nm). Contrary to this, the SCiO covers a small range of the far-red end of the visible spectrum.

Performance of manufacturer's software compared to The Unscrambler®

Various previous studies tried to determine the performance of the three food-scanners used in this study for the prediction of internal quality traits of selected fruit and vegetables. However, most studies only used the portable NIR spectrometers for collection of spectra and utilized the corresponding software tools provided by manufacturers for data extraction. In many scientific studies, the subsequent multivariate data analysis was performed by using specialized software for multivariate data analysis such as MATLAB (KAUR et al., 2017; TEYE et al., 2019) or The Unscrambler® (NCAMA et al., 2018). To the best of our knowledge, a direct comparison of these provided software tools to specialized software has only been done by LI et al. (2018) for the SCiO™ device with three kiwi quality parameters during a quality estimation trial. The authors obtained similar predictive performance for dry matter, TSS and firmness using SCiO™ Lab as well as The Unscrambler®. However, such a comparison has not yet been conducted for the F-750, which also provides its own software for evaluation and building of prediction models. In order to evaluate the potential application of these portable food-scanners in practice along the fresh fruit supply chain, this study used manufacturer's software as well as The Unscrambler® software for building of prediction models for a variety of tomato fruit quality traits. By comparing the results derived from both software tools, the accuracy and reliability of prediction models computed with manufacturer's software can be assessed.

A direct comparison of the prediction models created with The Unscrambler® (Table 3) to models developed using manufacturer's software (Table 4) indicates similar predictive performances for models of high accuracy ($r^2_{CV} > 0.90$) for both devices. The small variations in model performance was most likely due to slightly deviating numbers of principal components (PC), which are selected automatically and individually by each respective software. Additionally, the variability in sampling during cross-validation could cause these minor differences. Some prediction models of poor to moderate accuracy ($0.40 < r^2_{CV} < 0.80$) achieved similar results for both The Unscrambler® and manufacturer's software (e.g., color values b* and C* for the F-750 as well as color value C* for SCiO™ within both parts of the experiment). However, most prediction models within this area of accuracy computed with The Unscrambler® surpass those derived from F-750 or SCiO™ manufacturer's evaluation software. Whereas the evaluation of DM and acidity with The Unscrambler® within the first part of the experiment indicated merely moderate predictive capabilities, neither F-750 or SCiO™ software obtained any form of linear correlation between food-scanner spectra and reference values.

It should be noted that the evaluation of spectra using manufacturer's software yielded similar results to evaluations applying The Unscrambler® software for prediction models of high accuracy ($r^2_{CV} > 0.90$). In particular, these models of high accuracy are relevant when it comes to the practical application of food-scanners. This study therefore was able to demonstrate that existing food-scanners, including the software supplied, are suitable for creating feasible prediction models for fruit quality traits. Therefore, actors along the fresh fruit supply chain can use these devices independently and are not reliant on additional professional software for data evaluation. These results are in line with findings of LI et al. (2018), who found similar model accuracies for kiwi quality attributes using both The Unscrambler® and SCiO™ software. The results of this study demonstrate the great predictive potential and informative value of commercially available portable food-scanners. By combining multiple prediction models of high accuracy of various quality traits within one global prediction model, portable food-scanners can be used as multidimensional and non-destructive measurement tools of fruit quality. Therefore, future users can gain a comprehensive picture of fruit quality using only one single scan.

This study used two different experimental approaches in order to generate high variability of tomato fruit quality parameters. The utilization of fruit during ripening and storage resulted in a larger range of reference values L*, a*, b*, C*, h° and firmness compared to fruit from retail stores. Vice versa, differences in variety and origin within the second experiment increased variability of reference values DM, TSS and acidity. With the exception of color value b*, NIR prediction models yielded higher correlation for the experiment with a larger range of reference values. When it comes to the development of new prediction models by users in daily practice, a broad range of fruit qualities and respective reference values must be taken into account in order to obtain good prediction models. As highlighted in previous research (FAN et al., 2019; WEDDING et al., 2013), additional integration of biological and seasonal variability of fruit can contribute to build more reliable and robust prediction models and thereby unlock potential of non-destructive quality assessment via NIR food-scanners in practical applications.

Comparison of usability characteristics

A direct comparison of the three food-scanners used in this study shows clear differences with regard to usability and applicability characteristics (Table 5).

Both F-750 and H-100F are not reliant on internet access during scanning and can therefore be used independently at all points of the fresh produce supply chain, from orchards to incoming goods control in warehouses to retail stores. In contrast, the SCiO™ requires constant internet access during scanning as well as an additional mobile device (e.g., smartphone, tablet). This finding is contrary to statements on the SCiO™ website, where offline scanning is advertised (CONSUMER PHYSICS, 2020). However, personal experience as well as communication with SCiO™'s manufacturer revealed that offline scanning is not possible (CONSUMER PHYSICS, 2017). Therefore, the SCiO™ can not be used as an independent measuring device. Additionally, the device can only be used effectively in places with a stable internet connection, which is not always guaranteed in large warehouses or remote orchards.

Data handling of the F-750 can be described as very user-friendly. On the one hand, spectra are provided freely, on the other hand the gratis software and instructions for model building allows non-experts of NIR spectroscopy the building of new prediction models. With respect to the practical use of food-scanners for fresh produce quality control, new quality parameters could be developed by users along the supply chain themselves. In contrast, users of the SCiO™ device need to purchase an additional license to access raw spectra for building of own prediction models. To enable the use of these models for future predictions, further know-how in app programming is required. For the H-100F, spectra are also freely accessible, but the manufacturer does not provide software for building of own prediction models. According to the manufacturer, a short form of model building program was provided in the past, but deemed not useful. In order to analyze the spectra and respective reference data and produce calibration models in a suitable data format with the H-100F, model building programs such as The Unscrambler® can be used (SUNFOREST, 2018). However, these specialized programs are not suitable for many users in day-to-day practice, since on the on hand they are very expensive and on the other hand they require a lot of training.

The evaluation of the practical handling of all three food-scanners showed comparable effort in device maintenance. Additionally, none of these devices requires further running costs. Both F-750 and H-100F are characterized by very long battery runtimes and can therefore be used for several days in practical applications. The battery of the SCiO™ lasts for approximately 250 measurements. When used continuously, e.g., during quality assessment in the orchard or incoming goods control, this number can be reached within a short period of time. Theoretically, the battery life of the SCiO™ can be extended by using a mobile powerbank and charging the device parallel to scanning. However, an additional device and thus another dependency would be necessary. An important aspect in practical handling is the scan each device requires for one measurement, since it serves as indication for labor efficiency. Previous work identified rapid measurements as important feature of portable food-scanners (GOISSER et al., 2020a). Our measurement results for scan speed (Table 5) surpass those reported in previous studies (KAUR et al., 2017) as well as manufacturer's specifications (SUNFOREST, 2020; CONSUMER PHYSICS, 2020). It should be noted that in contrast to these references, we recorded the whole time interval from the start of one scan to the beginning of the next possible scan. For the SCiO™, quality and stability of the internet connection is essential with regard to scan speed. Both F-750 and H-100F display the scan result a few seconds before the next scan is possible, which most likely explains these deviations from manufacturer's specifications. However, the scan speed measured in our study is much more practice-oriented. Assuming a fixed timeframe for the application of food-scanners in quality control processes, different numbers of scans can be performed with each device. Depending on the device, different numbers of measuring points would be available for quality assessment. Due to the built-in technology, weight and dimension differs notably between devices. These features have to be considered when the devices are used portable in the orchard or warehouse and have to be physically carried around by the users.

All manufacturers provide accessories that enable the measurement of fruit and vegetables of various sizes. Therefore, users along the fresh produce supply chain are not limited in their use of these devices due to fruit size. However, additional fruit properties such as the color and thickness of fruit skin have to be considered when applying NIR spectroscopy, since these properties could greatly influence the accuracy and reliability of NIR prediction models. The F-750 provides a GPS function, which enables the measurement values to be displayed in a map view. By using RFID tags in orchards in combination with the built-in RFID reader, the H-100F takes a similar approach. On the one hand, fruit ripeness can be monitored in orchards, on the other hand, this function can also be used to verify the origin of produce as well as quality of origin by measurements at production companies. Additional accessories for the measurement of liquids can be purchased for the SCiO, the F-750 provides a purchasable liquids and powders kit. Users are therefore able to use these portable NIR instruments for testing various substances within the food industry (e.g., fruit juices, beverages, dried herbs) or completely different areas of application in other industries.

Conclusion

The predictability of various tomato quality parameters using three commercially available portable food-scanners was analyzed and the software for building of own prediction models provided by manufacturers of these devices evaluated through comparison to state-of-the-art software for multivariate analysis. The results highlight the capability of these devices in predicting various internal fruit quality parameters in a non-destructive way. Our findings are consistent with previous research, which in many cases used laboratory NIR instruments compared to portable and miniaturized NIR devices. Quality control of fruit and vegetable along the fresh produce supply chain is often limited to optical inspections and in some cases not performed at all. Through the combination of a multitude of internal quality measurements within a single scan, food-scanners provide a comprehensive and objective picture of fruit quality. Food-scanners can be used on various points along the fresh produce supply chain and replace traditional time-consuming and elaborate destructive measurement methods of internal fruit quality. Therefore, food-scanners can be an important tool in making fruit quality along the whole fresh produce supply chain more transparent and comprehensible.

The comparison of software provided by manufacturer to state-of-the-art software shows that the provided software delivers good results, especially for models of high prediction accuracy. Applied to the practical application of these devices in day-to-day processes of quality control this means that special expertise in the field of NIR spectroscopy and model building is not necessary. Since some manufacturers provide basic software solutions in combination with operating instructions, users are enabled to build their own prediction models.

As highlighted in this study, variation in fruit quality is of great importance in order to build accurate prediction models. The variation of the respective fruit quality parameter can be influenced in various ways, e.g., through different ripening stages, variation in origin and cultivars or seasonal differences. With regard to the practical use and long term application of food-scanners along the fresh produce supply chain, it is vital that users of these devices are aware of the importance of the variability of reference values when it comes to creating new, reliable and robust prediction models.

Based on previous studies and personal experience with supply chain actors, an exemplary framework evaluating food-scanner characteristics with respect to usability and applicability was designed. The evaluation of these usability characteristics showed various differences between the three devices used in this study, especially with regard to the independent use of these devices and handling of generated data. Potential users of food-scanners must inform themselves in advance and make sure that the respective device meets their company's requirements, especially in day-to-day work processes of fruit quality control. For a further spread and effective use of these devices in practice, manufacturers must pay attention to user-friendliness and simple model building solutions.

Acknowledgements

The research was financed by the Bavarian Ministry of Food, Agriculture and Forestry as part of the alliance "Wir retten Lebensmittel" [We Save Foodstuffs] as well as the QS Science Funds in Fruit, Vegetables and Potatoes. The authors address special thanks to STEP Systems GmbH from Nuremberg, Germany for their loan of two food-scanner devices during the course of this experiment.

Conflict of Interest

No potential conflict of interest was reported by the authors

Tables

Tab. 1: Sample composition of supermarket tomatoes.

Tomato type	Variety	N	Origin
Salad	n/a	25	The Netherlands
Salad	n/a	25	The Netherlands
Salad	Prince	25	Germany
Roma	n/a	25	Spain
Miniature Roma	Aromatica	25	The Netherlands
Miniature Roma	Sunstream	25	The Netherlands
Cherry	Romantic	25	Spain
Cherry	n/a	25	Germany

Tab. 2: Distribution of reference values of quality parameters within each part of the experiment. L*, a*, b*, C*, h°: colorimetric values; DM: dry matter; TSS: total soluble solids; SD: standard deviation

Parameter	Experiment part	N	Range	Mean	SD
L*	1	245	29.13 - 51.36	36.03	5.40
	2	120	32.33 - 41.88	36.12	2.11
a*	1	245	-7.97 - 31.69	17.08	13.19
	2	120	11.55 - 28.46	22.29	2.87
b*	1	245	18.01 - 39.27	27.65	3.34
	2	120	16.49 - 27.99	21.16	2.57
C*	1	245	21.49 - 49.26	34.74	6.11
	2	120	23.66 - 39.59	30.85	2.79
h°	1	245	40.87 - 106.54	62.07	22.50
	2	120	35.18 - 63.39	43.56	5.02
Firmness (N)	1	245	6.14 - 90.75	30.53	19.26
	2	120	12.67 - 42.24	24.43	7.21
DM (%)	1	80	5.50 - 7.58	6.09	0.38
	2	80	4.02 - 9.69	6.72	1.74
TSS (%)	1	165	4.50 - 7.30	5.54	0.59
	2	120	3.1 - 9.00	5.78	1.84
Acidity (g/L)	1	165	2.43 - 6.02	3.73	0.77
	2	120	2.40 - 9.21	5.08	1.32
Brix/Acid ratio	1	165	8.63 - 27.21	15.60	3.98
	2	120	7.38 - 18.22	11.42	2.39

Tab. 3: Prediction of tomato quality parameters of the F-750 Produce Quality Meter and SCiO™ Molecular Sensor using the respective manufacturer's software. L*, a*, b*, C*, h°: colorimetric values; TSS: total soluble solids; r^2_C: r^2 of calibration; $RMSE_C$: root mean square error of calibration; r^2_{CV}: r^2 of cross-validation; $RMSE_{CV}$: root mean square error of cross validation; PC: principal components

			F-750 (λ = 477 - 1059 nm)					SCiO™ (λ = 740 - 1070 nm)		
Parameter	Experiment part	N	r^2_C	$RMSE_C$	r^2_{CV}	$RMSE_{CV}$	PC	r^2_P	SEP	PC
*L**	1	245	0.90	1.68	0.90	1.70	3	0.89	1.83	9
	2	120	0.54	1.44	0.52	1.46	3	0.47	1.54	3
*a**	1	245	0.95	2.83	0.95	2.88	4	0.92	3.81	12
	2	120	0.21	2.54	0.19	2.58	3	0.37	2.28	3
*b**	1	245	0.27	2.85	0.25	2.89	3	0.42	2.56	5
	2	120	0.66	1.49	0.65	1.52	3	0.72	1.36	4
*C**	1	245	0.74	3.15	0.72	3.24	4	0.80	2.76	7
	2	120	0.41	2.14	0.39	2.17	2	0.62	1.73	4
h°	1	245	0.96	4.60	0.96	4.71	4	0.91	6.95	12
	2	120	0.42	3.84	0.40	3.90	3	0.39	3.91	2
Firmness (N)	1	245	0.92	5.40	0.92	5.53	4	0.91	5.84	10
	2	120	0.49	5.16	0.47	5.25	3	0.46	5.29	6
DM (%)	1	80	0.08	0.59	0.03	0.61	1	0.07	0.37	1
	2	80	0.94	0.41	0.93	0.47	6	0.97	0.31	8
TSS (%)	1	165	0.56	0.39	0.56	0.39	2	0.71	0.32	7
	2	120	0.93	0.47	0.92	0.51	6	0.97	0.30	8
Acidity (g/L)	1	165	0.03	2.00	0.02	2.01	2	0.00	2.00	1
	2	120	0.51	0.92	0.49	0.94	3	0.66	0.77	6
Brix/Acid ratio	1	165	0.43	3.09	0.42	3.13	2	0.49	2.92	4
	2	120	0.46	1.76	0.46	1.77	2	0.51	1.68	3

Tab. 4: Prediction of tomato quality parameters of three commercially available NIR devices using The Unscrambler® software for evaluation. L*, a*, b*, C*, h°: colorimetric values; TSS: total soluble solids; r^2_C: r^2 of calibration; $RMSE_C$: root mean square error of calibration; r^2_{CV}: r^2 of cross-validation; $RMSE_{CV}$: root mean square error of cross validation; PC: principal components

			F-750 (λ = 477 - 1059 nm)					SCiO™ (λ = 740 - 1070 nm)					H-100F (λ = 650 - 950 nm)				
Parameter	Experiment	N	r^2_C	$RMSE_C$	r^2_{CV}	$RMSE_{CV}$	PC	r^2_C	$RMSE_C$	r^2_{CV}	$RMSE_{CV}$	PC	r^2_C	$RMSE_C$	r^2_{CV}	$RMSE_{CV}$	PC
*L**	1	245	0.90	1.70	0.90	1.74	1	0.91	1.61	0.90	1.74	9	0.93	1.39	0.93	1.44	3
	2	120	0.71	1.14	0.62	1.32	6	0.62	1.31	0.57	1.40	4	0.75	1.05	0.70	1.16	6
*a**	1	245	0.97	2.44	0.96	2.53	4	0.93	3.41	0.92	3.71	9	0.96	2.65	0.96	2.74	3
	2	120	0.61	1.78	0.48	2.08	6	0.55	1.91	0.49	2.05	5	0.63	1.74	0.54	1.95	7
*b**	1	245	0.31	2.78	0.29	2.83	3	0.60	2.12	0.55	2.26	8	0.53	2.30	0.48	2.43	5
	2	120	0.68	1.45	0.66	1.51	3	0.76	1.25	0.73	1.35	4	0.78	1.22	0.72	1.38	7
*C**	1	245	0.75	3.04	0.73	3.20	4	0.85	2.36	0.83	2.54	8	0.82	2.58	0.80	2.71	5
	2	120	0.43	2.10	0.42	2.14	1	0.69	1.56	0.65	1.67	5	0.60	1.78	0.49	2.00	7
h°	1	245	0.96	4.52	0.96	4.70	3	0.93	6.04	0.92	6.50	9	0.95	4.86	0.95	4.95	2
	2	120	0.72	2.66	0.64	3.03	6	0.61	3.15	0.54	3.45	5	0.78	2.37	0.73	2.64	6
Firmness (N)	1	245	0.93	5.17	0.93	5.31	3	0.90	6.07	0.89	6.55	9	0.92	5.35	0.92	5.46	2
	2	120	0.73	3.77	0.64	4.37	6	0.66	4.19	0.54	4.95	11	0.72	3.80	0.67	4.18	6
DM (%)	1	80	0.47	1.58	0.18	1.99	7	0.54	0.28	0.32	0.33	6	0.33	0.32	0.16	0.36	5
	2	80	0.96	0.33	0.94	0.43	7	0.98	0.25	0.97	0.32	8	0.96	0.35	0.94	0.42	7
TSS (%)	1	165	0.80	0.26	0.69	0.32	8	0.85	0.23	0.81	0.26	10	0.77	0.28	0.67	0.34	11
	2	120	0.94	0.46	0.92	0.51	5	0.97	0.32	0.96	0.35	6	0.97	0.32	0.96	0.38	7
Acidity (g/L)	1	165	0.60	0.49	0.51	0.54	6	0.63	0.46	0.57	0.51	8	0.68	0.43	0.63	0.47	5
	2	120	0.74	0.67	0.64	0.79	6	0.71	0.71	0.66	0.78	6	0.82	0.57	0.75	0.67	8
Brix/Acid ratio	1	165	0.74	2.11	0.69	2.30	6	0.75	2.06	0.70	2.25	7	0.77	1.96	0.74	2.10	6
	2	120	0.62	1.47	0.54	1.64	5	0.77	1.15	0.65	1.42	12	0.71	1.29	0.64	1.45	7

Table 5: Compilation of food-scanner characteristics with respect to usability and applicability

Features	Device		
	F-750	SCiO™	H-100F
Device independency			
Necessity of internet access during scanning	No	Yes	No
Additional devices needed for scanning	None	Smartphone / Tablet	None
Internal / external white-referencing	Internal	External	External
Data handling			
Access of spectra	Free, SD-Card	Licenced, online database	Free, USB, Bluetooth
Ability for building own prediction models	Free software available	Programming of own apps necessary	No
Specialized knowledge necessary for model building	No	Yes	Yes
Practical handling			
Instrument maintenance	Lens cleaning	Lens and LED cleaning	Lens and light source cleaning
Running costs	None	None	None
Durability of battery life	> 1600 scans[a]	> 250 scans[a]	~ 2000 scans[b]
Scan speed (mean and standard deviation)	14.3 ± 0.1 s[b]	8.2 ± 0.7 s[b]	4.9 ± 0.1 s[b]
Device weight	1.05 kg	36 g	440 g
Device dimension	18 x 13.5 x 6 cm	5.5 x 3.7 x 1.4 cm	19 x 17 x 10.7 cm
Available accessories			
Provided accessories	Small fruit adaptor, GPS	Small object holder	Rubber mold for small fruit, built-in RFID reader
Purchasable accessories	Liquids and powders kit	Liquid accessory	n/s

[a]According to manufacturer's specifications

[b]Results of own measurements

References

Abbott, J.A., 1999: Quality measurement of fruits and vegetables. Postharvest Biology and Technology 15, 207-225. DOI: 10.1016/S0925-5214(98)00086-6.

Batu, A., 2004: Determination of acceptable firmness and colour values of tomatoes. Journal of Food Engineering 61, 471-475. DOI: 10.1016/S0260-8774(03)00141-9.

BLE, 2019: Obst und Gemüse von A bis Z. Vermarktungsnormen frisches Obst und Gemüse. Retrieved from https://www.ble.de/DE/Themen/Ernaehrung-Lebensmittel/Vermarktungsnormen/Obst-Gemuese/Vermarktungsnormen-Hilfen-zur-Anwendung/ObstundGemueseA_Z.html?nn=8904900.

Bruhn, C.M., Feldman, N., Garlitz, C., Harwood, J., Ivans, E., Marshall, M., Riley, A., Thurber, D., Williamson, E., 1991: Consumer perception of quality: apricots, cantaloupes, peaches, pears, strawberries, and tomatoes. Journal of Food Quality 14, 187-195. DOI: 10.1111/j.1745-4557.1991.tb00060.x.

Casals, J., Rivera, A., Sabaté, J., Romero del Castillo, R., Simó, J., 2019: Cherry and Fresh Market Tomatoes: Differences in Chemical, Morphological, and Sensory Traits and Their Implications for Consumer Acceptance. Agronomy 9, 9. DOI: 10.3390/agronomy9010009.

Causse, M., Damidaux, R., Rousselle, P., 2007: Traditional and Enhanced Breeding for Quality Traits in Tomato, 153-192. In: Razdan, M.K. (ed.), Genetic improvement of solanaceous crops. Enfield, NH: Science Publ.

Clément, A., Dorais, M., Vernon, M., 2008: Nondestructive measurement of fresh tomato lycopene content and other physicochemical characteristics using visible-NIR spectroscopy. Journal of agricultural and food chemistry 56, 9813-9818. DOI: 10.1021/jf801299r.

Consumer Physics, 2017: Customer Support. Persönliche Mitteilung, Freising, Germany.

Consumer Physics, 2020: Technology. Technical Specifications. Retrieved 28 April 2020, from https://www.consumerphysics.com/technology/.

Dos Santos, C.A.T., Lopo, M., Páscoa, R.N.M.J., Lopes, J.A., 2013: A review on the applications of portable near-infrared spectrometers in the agro-food industry. Applied spectroscopy 67, 1215-1233. DOI: 10.1366/13-07228.

Ecarnot, M., Baczyk, P., Tessarotto, L., Chervin, C., 2013: Rapid phenotyping of the tomato fruit model, Micro-Tom, with a portable VIS-NIR spectrometer. Plant physiology and biochemistry: PPB 70, 159-163. DOI: 10.1016/j.plaphy.2013.05.019.

Ercolano, M.R., Carli, P., Soria, A., Cascone, A., Fogliano, V., Frusciante, L., Barone, A., 2008: Biochemical, sensorial and genomic profiling of traditional Italian tomato varieties. Euphytica 164, 571-582. DOI: 10.1007/s10681-008-9768-4.

Fan, S., Li, J., Xia, Y., Tian, X., Guo, Z., Huang, W., 2019: Long-term evaluation of soluble solids content of apples with biological variability by using near-infrared spectroscopy and calibration transfer method. Postharvest Biology and Technology 151, 79-87. DOI: 10.1016/j.postharvbio.2019.02.001.

Flores, K., Sánchez, M.-T., Pérez-Marín, D., Guerrero, J.-E., Garrido-Varo, A., 2009: Feasibility in NIRS instruments for predicting internal quality in intact tomato. Journal of Food Engineering 91, 311-318. DOI: 10.1016/j.jfoodeng.2008.09.013.

Folta, K.M., Klee, H.J., 2016: Sensory sacrifices when we mass-produce mass produce. Horticulture research 3, 16032. DOI: 10.1038/hortres.2016.32.

Goisser, S., Mempel, H., Bitsch, V., 2020a: Food-Scanners as a Radical Innovation in German Fresh Produce Supply Chains. International Journal on Food System Dynamics, Vol 11, No 2 (2020) / International Journal on Food System Dynamics, Vol 11, No 2 (2020) 11, 101-116. DOI: 10.18461/IJFSD.V11I2.43.

Goisser, S., Wittmann, S., Fernandes, M., Mempel, H., Ulrichs, C., 2020b: Comparison of colorimeter and different portable food-scanners for non-destructive prediction of lycopene content in tomato fruit. Postharvest Biology and Technology 167, 111232. DOI: 10.1016/j.postharvbio.2020.111232.

He, Y., Zhang, Y., Pereira, A.G., Gomez, A.H., Wang, J., 2005: Nondestructive Determination of Tomato Fruit Quality Characteristics Using Vis/NIR Spectroscopy Technique. International Journal of Information Technology 11, 97-108.

Jha, S.N., Matsuoka, T., 2004: Non-destructive determination of acid-brix ratio of tomato juice using near infrared spectroscopy. Int J Food Sci Tech 39, 425-430. DOI: 10.1111/j.1365-2621.2004.00800.x.

Kaur, H., Künnemeyer, R., McGlone, A., 2017: Comparison of hand-held near infrared spectrophotometers for fruit dry matter assessment. Journal of Near Infrared Spectroscopy 25, 267-277. DOI: 10.1177/0967033517725530.

Kusumiyati, Akinaga, T., Tanaka, M., Kawasaki, S., 2008: On-tree and after-harvesting evaluation of firmness, color and lycopene content of tomato fruit using portable NIR spectroscopy. Journal of Food, Agriculture & Environment 6, 327-332.

Li, M., Qian, Z., Shi, B., Medlicott, J., East, A., 2018: Evaluating the performance of a consumer scale SCiO™ molecular sensor to predict quality of horticultural products. Postharvest Biology and Technology 145, 183-192. DOI: 10.1016/j.postharvbio.2018.07.009.

McGlone, V.A., Kawano, S., 1998: Firmness, dry-matter and soluble-solids assessment of postharvest kiwifruit by NIR spectroscopy. Postharvest Biology and Technology 13, 131-141. DOI: 10.1016/S0925-5214(98)00007-6.

Ncama, K., Magwaza, L.S., Poblete-Echeverría, C.A., Nieuwoudt, H.H., Tesfay, S.Z., Mditshwa, A., 2018: On-tree indexing of 'Hass' avocado fruit by non-destructive assessment of pulp dry matter and oil content. Biosystems Engineering 174, 41-49. DOI: 10.1016/j.biosystemseng.2018.06.011.

OECD, 2018: OECD Fruit and Vegetable Scheme. Guidelines on objective tests to determine quality of fruit and vegetables, dry and dried produce. Retrieved from www.oecd.org/agriculture/fruit-vegetables/.

Oliveira, G.A. de, Bureau, S., Renard, C.M.-G.C., Pereira-Netto, A.B., Castilhos, F. de, 2014: Comparison of NIRS approach for prediction of internal quality traits in three fruit species. Food chemistry 143, 223-230. DOI: 10.1016/j.foodchem.2013.07.122.

Pasquini, C., 2003: Near Infrared Spectroscopy: Fundamentals, Practical Aspects and Analytical Applications. Journal of the Brazilian Chemical Society 14, 198-219. DOI: 10.1590/S0103-50532003000200006.

Rateni, G., Dario, P., Cavallo, F., 2017: Smartphone-Based Food Diagnostic Technologies: A Review. Sensors (Basel, Switzerland) 17. DOI: 10.3390/s17061453.

Saad, A., Jha, S.N., Jaiswal, P., Srivastava, N., Helyes, L., 2016: Non-destructive quality monitoring of stored tomatoes using VIS-NIR spectroscopy. Engineering in Agriculture, Environment and Food 9, 158-164. DOI: 10.1016/j.eaef.2015.10.004.

Sheng, R., Cheng, W., Li, H., Ali, S., Akomeah Agyekum, A., Chen, Q., 2019: Model development for soluble solids and lycopene contents of cherry tomato at different temperatures using near-infrared spectroscopy. Postharvest Biology and Technology 156, 110952. DOI: 10.1016/j.postharvbio.2019.110952.

Statista, 2019: Gemüsekonsum in Deutschland. Retrieved from https://de.statista.com/statistik/studie/id/42820/dokument/gemuesekonsum-in-deutschland/.

Su, H., Sha, K., Zhang, L., Zhang, Q., Xu, Y., Zhang, R., Li, H., Sun, B., 2014: Development of near infrared reflectance spectroscopy to predict chemical composition with a wide range of variability in beef. Meat science 98, 110-114. DOI: 10.1016/j.meatsci.2013.12.019.

Sunforest, 2018: Application of Sunforest H-100F. Persönliche Mitteilung, Freising, Germany.

Sunforest, 2020: H-100F. Portable Nondestructive Fruit Quality Meter. Retrieved 29 April 2020, from http://sunforest.kr/category_main.php?sm_idx=169.

Teye, E., Amuah, C.L.Y., McGrath, T., Elliott, C., 2019: Innovative and rapid analysis for rice authenticity using hand-held NIR spectrometry and chemometrics. Spectrochimica acta. Part A, Molecular and biomolecular spectroscopy 217, 147-154. DOI: 10.1016/j.saa.2019.03.085.

Torres, I., Pérez-Marín, D., La Haba, M.-J.D., Sánchez, M.-T., 2015: Fast and accurate quality assessment of Raf tomatoes using NIRS technology. Postharvest Biology and Technology 107, 9-15. DOI: 10.1016/j.postharvbio.2015.04.004.

UNECE, 2018: UNECE Standard FFV-36 concerning the marketing and commercial quality control of tomatoes. 2017 Edition. Retrieved from https://www.unece.org/trade/agr/standard/fresh/ffv-standardse.html.

USDA, 1975: USDA Color Chart: Color Classification Requirements in Tomatoes. Retrieved from https://ucanr.edu/repository/view.cfm?article=83755%20&groupid=9.

Wedding, B.B., Wright, C., Grauf, S., White, R.D., Tilse, B., Gadek, P., 2013: Effects of seasonal variability on FT-NIR prediction of dry matter content for whole Hass avocado fruit. Postharvest Biology and Technology 75, 9-16. DOI: 10.1016/j.postharvbio.2012.04.016.

ORCID

Simon Goisser: 0000-0002-7685-8893

Sabine Wittmann: 0000-0002-8669-4132

Christian Ulrichs: 0000-0002-6733-3366

Heike Mempel: 0000-0002-5820-663X

Address of the corresponding author

Heike Mempel, Greenhouse Technology and Quality Management, University of Applied Sciences Weihenstephan-Triesdorf, Am Staudengarten 10, 85354 Freising, Germany
E-mail: heike.mempel@hswt.de

6 Determination of tomato quality attributes using portable NIR-sensors

In: OCM 2019 - 4th International Conference on Optical Characterization of Materials, p. 1-12

Cite as: Goisser, S.; Krause, J.; Fernandes, M.; Mempel, H. (2019): Determination of tomato quality attributes using portable NIR-sensors. In: Jürgen Beyerer, Fernando Puente León und Thomas Längle (Hg.): OCM 2019 - 4th International Conference on Optical Characterization of Materials. March 13th - 14th, 2019. Karlsruhe: Scientific Publishing, p. 1–12.

DOI: https://doi.org/10.5445/KSP/1000087509

Determination of tomato quality attributes using portable NIR-sensors

Simon Goisser[1], Julius Krause[2,3], Michael Fernandes[4], and Heike Mempel[1]

[1] University of Applied Sciences Weihenstephan Triesdorf, Greenhouse Technology and Quality Management, Am Staudengarten 10, 85354 Freising
[2] Karlsruhe Institute of Technology, Vision and Fusion Laboratory IES, Haid-und-Neu-Str. 7, 76131 Karlsruhe
[3] Fraunhofer Institute of Optronics, System Technologies and Image Exploitation IOSB, Visual Inspection Systems, Fraunhoferstr. 1, 76131 Karlsruhe
[4] Deggendorf Institute of Technology, Technology Campus Grafenau, Department of Applied Artificial Intelligence, Hauptstraße 3, 94481 Grafenau

Abstract As part of a research project a multidisciplinary approach of different research institutes is followed to investigate the possibility of using a commercially available miniaturized NIR-sensor for the determination of tomato fruit quality parameters in postharvest. Correlation of spectra and tomato reference values of firmness, dry matter and total soluble solids showed good prediction accuracy. Additionally the decline of firmness over storage time with respect to storage temperature of tomatoes could be modelled. Therefore, the decline of firmness as an indicator for shelf-life can be predicted using this portable NIR-Sensor.

Keywords: NIR, portable, tomato, brix, firmness, dry matter.

1 Introduction

At the moment, grading and sorting of fresh produce is highly dominated by external and internal quality attributes like colour, fruit texture and sugar content. Some of these requirements are statutory and written down in marketing standards for fresh fruit and vegetables [1]. In order to guarantee internal quality like sweetness and taste of produce, additional testing of internal quality parameters like sugar content or sugar-acid ratio is necessary for certain products alongside these statutory criteria standards. The determination of internal parameters is often time consuming and requires destructive measurement methods. Immediately after harvest the quality of fruits and vegetables changes due to ongoing metabolic processes. Depending on various parameters like fruit maturity, packaging and storage conditions, quality of post harvest produce decays in different time periods. Sensitive fruits like strawberries have a shelf life of only a few days, whereas more robust fruits like apples can be stored for up to nearly one year under appropriate storage conditions. Climacteric fruits like tomatoes underlie post-ripening, which on one hand can lead to longer shelf life, but on the other hand cause alteration of sensorial parameters like taste or haptic. Furthermore, firmness is an important indicator of tomato quality which determines shelf life and influences consumer's acceptance [2]. Tomato is one of the most important fruits cultivated in Germany, with a total percentage of nearly 30 % of Germany's greenhouse production area for vegetables [3] and number one vegetable with respect to per capita consumption [4]. Sugar content, acidity and the acid-brix ratio are internal quality parameters that can help to determine the ripening stage and affect the taste of tomatoes. In the past, various studies were conducted to prove that some of these parameters can be predicted using near-infrared spectroscopy (NIRS) on different tomato varieties [5, 6]. NIRS is well suited for quality control of fresh produce because it is non-destructive and requires little to no sample preparation. Additionally, NIRS techniques can be used as multidimensional predictors to determine various parameters in one work-step. Due to an ongoing technical development and miniaturization in the field of NIR spectrometers, companies are offering small and portable sensors. These devices can be used in numerous applications throughout the agro-food and horticultural industry, as illustrated by dos Santos et al. [7]. The technique of NIR is already applied in sorting and grading machines, especially for high quality produce which are ripened in post-harvest processes (e. g., mango, avocado). In contrast to bulky benchtop devices, these hand-held sensors allow a transfer of NIR techniques from the laboratory to in-field and other applications along the whole horticultural supply chain of fresh produce. In addition to the determination of specific fruit quality parameters, these devices could unlock potential in measuring the maturity or ripening stage of fresh produce with respect to remaining shelf life. A measurement tool which allows the quantification of shelf life could help to reduce the amount of annual food loss of around 11 million tons [8] in Germany. First studies indicate that there is high potential for predicting the quality of fresh fruits and vegetables with portable and miniaturized NIR devices. According to Kusumiyati et al. [9], the use of a portable NIR-device and PLSR analysis proved feasibility of predicting the on-tree firmness of tomatoes with r^2 = 0,88 and a standard error of prediction of 0,09 MPa as well as lightness value L* (r^2 = 0.96, SEP = 3.19) and color value a* (r^2 = 0,98, SEP = 3,13). Some startup companies focus on this development and launch various miniaturized NIR-handheld devices called food-scanners, promising end-consumers a fast and nondestructive measurement of various food traits like protein, sugar or total energy content [10]. A first study by Kaur et al. [11], which examined the performance of different portable and commercially available spectrometers with respect to the prediction of dry matter, showed promising results for a combined data set of apples, kiwifruit and summerfruit (rp^2 of 0,93-0,95). Subsequent research investigated the on-tree prediction of 'Hass' avocado harvest maturity using the F-750 spectrometer (Felix Instruments) and found suitable prediction models for dry matter (rp^2 = 0.98; RMSEP = 0.25 %) and oil content (rp^2 = 0.96; RMSEP = 1.14 %) [12]. A current study, focusing on the performance of a consumer scale SCiO (Consumer Physics) molecular sensor by Li et al. (2018), found good results for the prediction of total soluble solids in kiwifruit ($RVal^2$ = 0.77; RMSEp = 0.76 %) and potential in classifying feijoa according to maturity and 'Hass' avocado according to ripening stage, whereas the prediction of apple quality was not feasible. In summary, miniaturized NIR devices seem to hold potential in predicting the quality of fresh produce, making it a good method of choice in investigating the feasibility for quality predictions of tomato fruit. At the moment, a key challenge for a successful implementation in supply chain processes is the provision of suitable prediction models of fresh produce. Since different types of produce require specific calibrations, further work is required to investigate predictable fruit quality parameters and to build up data collections, which allow a robust calibration of portable devices. As part of the alliance "Wir retten Lebensmittel", initiated by the Bavarian Ministry of Food, Agriculture and Forestry, a multidisciplinary approach of different research institutes was conducted during a two-year research project. The aim of this work was to investigate the possibility of using a self-built and miniaturized NIR-sensor for the determination of tomato fruit maturity parameters such as sugar content, firmness and dry matter. A similar approach with focus on a commercially available pocket-sized NIR-sensor has already been performed (publication in review process). Based on these results, storage experiments were carried out in order to evaluate the feasibility of predicting shelf life of tomatoes with NIR-sensors. Tomato was chosen as model fruit because of its economic importance. An early determination of maturity as well as shelf life of tomatoes could help to control the supply chain and take alternative paths for produce not suitable for the fresh market (e. g., processing into soups, juices or smoothies). Furthermore, appropriate measures like sales promotions of ripe fruits can be launched in order to reduce food waste.

2 Material and methods

2.1 Sample material

Cherry- and salad-tomatoes (*Solanum lycopersicum* 'EZ 1359' and 'EZ 1256') from a greenhouse of the University of Applied Sciences Weihenstephan-Triesdorf (latitude 48°24′6″N and longitude 11°43′53″E) were used as sample material. Tomato plants were cultivated in a run-to-waste system on rock wool. Determination of sugar content, firmness, dry matter and color values of tomato fruits was done during the summer of 2017 at the Institute of Horticulture, Freising, Germany.

2.2 Methods and storage conditions

In a first experiments, 40 salad- and 40 cherry-tomatoes were harvested and the spectra recorded. Afterwards, sugar content in terms of TSS-concentration of each individual fruit was measured. In a second experiment, 120 tomatoes (60 cherry and 60 salad) were harvested and stored at room temperature (20 - 22 °C) for two weeks to evaluate post-ripening-processes with respect to changes in fruit skin color and dry matter. Every two days, spectra of ten fruits of each variety were taken and fresh weight, dry weight and color values of each fruit was recorded. In order to model shelf life and the decay of tomato fruit firmness over time, a third experiment was conducted using different storage conditions. 320 salad- and 360 cherry- tomatoes were harvested from the greenhouse. Spectra was recorded and firmness measured of 16 salad- and 20 cherry-tomatoes to determine initial fruit firmness. The remaining tomatoes were stored in batches of equal size under three different storage conditions at 8, 15 and 20 °C and relative humidity of 96 - 98 % respectively using laboratory refrigerators (Liebherr Mediline model LKPv 6522, Liebherr-International GmbH, Biberach, Germany). The resulting vapor pressure deficits were 2,1 - 4,3 hPa (8 °C), 3,4 - 6,8 hPa (15 °C) and 4,6 - 9,4 hPa (20°C). Every two to three days, the spectra of ten fruits of each variety and storage condition were recorded and firmness was measured. Due to moldiness and other physiological disorders during storage, 6 salad- and 11 cherry-tomatoes had to be excluded from the experiment, resulting in a valid data set of 314 salad- and 349 cherry-tomatoes.

2.3 Recording of NIR spectra

Spectroscopic measurements were performed using a self-built handheld NIR spectrometer consisting of a DLP® NIRscan™ Nano Evaluation Module (Texas Instruments, Dallas, Texas), supporting wavelengths from 900 - 1700 nm, embedded in a custom-built aluminum case (see Fig. 1A.). The method of measurement is diffuse reflection. Spectra of tomatoes were recorded by taking eight separate measurements orthogonally around the equator of each fruit (see 1.1) using a reprogrammed graphical user interface (GUI). The eight spectra for each fruit were averaged afterwards.

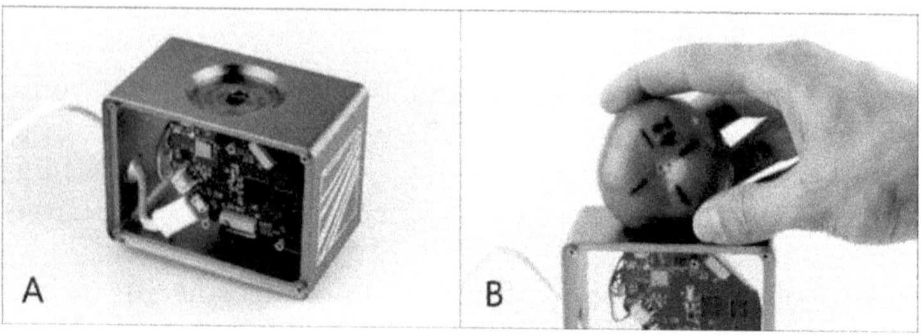

Figure 1.1: Custom-built hand held NIR spectrometer (A) and recording of tomato spectra around the fruit equator (B)

2.4 Acquisition of reference data

Reference measurements were performed immediately after recording spectra. Total soluble solids (TSS) were measured according to the OECD fruit and vegetable scheme by squeezing two longitudinal slices from opposite sides of the fruit with a garlic press and measuring the mixed juice with a digital refractometer HI 96801 (Hanna Instruments, Woonsocket, USA). The results, in degrees Brix, were recorded to one decimal place. The concentration of dry matter (DM) was measured gravimetrically for whole tomatoes. Weight of fresh fruits was determined to the nearest 0,001 g. Subsequently fruits were dried in an oven at 105 °C for 48 h. The final dry weight was determined to the nearest 0,001 g and DM calculated as the percentage of dry weight to initial fresh weight of each fruit. Firmness was measured using a non-invasive hand-held penetrometer AGROSTA 100X (Agro Technologie, Serqueux, France) specifically designed to measure tomatoes and berries. The penetrometer expresses firmness of fruits in a unit of percent in a range from 0 - 100, whereas 100 percent equals 8,09 Newton. Results were converted to Newton by taking account of penetrometer-pin-diameter and the maximum force used to push in the metal pin. Measurements were taken at two spots on opposite sides of the equator of each tomato. Both readings for each tomato were averaged. Additionally a subjective firmness bonitur was performed by the first author, grading tomatoes into the bonitur classes A (very firm), B (firm) and C (soft).

3 Results and discussion

The data analysis was performed with *Python* using the *scikit-learn* framework [13]. After a normalization in the first step, the regression models were developed to establish a relationship between the normalized spectra and the properties of salad and cherry tomatoes. To verify the models a *leave-one-out-cross-validation* (LOOCV) was performed.

3.1 Spectral preprocessing

To reduce scattering effects and noise, spectral preprocessing was applied. Because of the signal-to-noise ratio, the spectral range for the analysis was reduced to 198 bands between 901 nm and 1608 nm. For noise suppression, the mean value of all spectra of a sample was calculated and in addition, the mean value spectra were smoothed with a Savitzky-Golay filter (15,3). In a further step, the first derivative was formed to correct the baseline due to different scattering properties. Finally, different intensities were compensated by normalization using Standard Normal Variate (SNV).

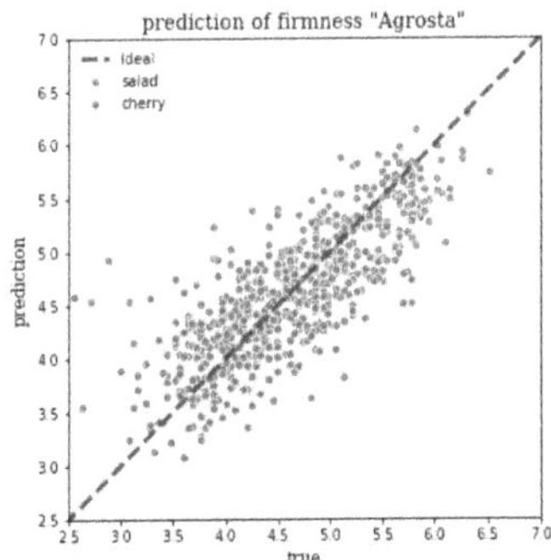

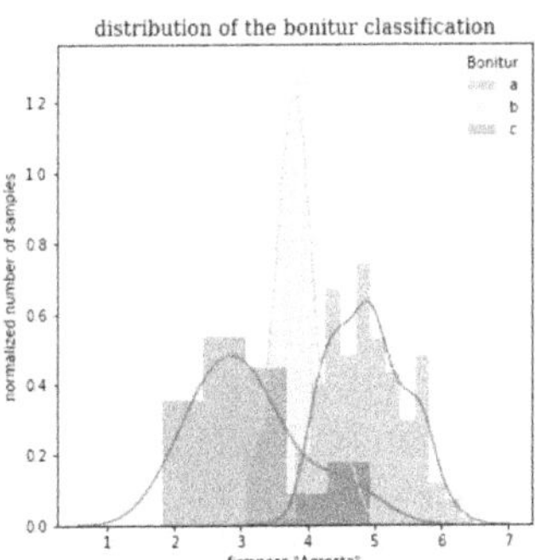

Figure 1.2: PLSR prediction of the dry matter using the normalized spectra. The multi-product model works well for both varieties. (RMESP 'salad' = 0.45, RMESP 'cocktail' =0.45). The bonitur classification is related to the firmness. Therefore, a clear dependence of the quality score on the measured firmness can be seen.

3.2 Prediction of quality attributes

The normalized spectra can be used to predict the firmness of the tomato by a partial least squares regression modell (PLSR) modell. A multi-product PLSR model was developed, which was trained with the spectra of both varieties. The root-mean-square-error-of-prediction (RMSEP) of the model determined in a cross validation is 0.52 for salad tomatoes and 0.5 for cherry tomatoes (see fig. 1.2).

There is also a relationship between the optically and mechanically measurable firmness and the quality evaluation of the bonitur (see fig. 1.2). This allows classification into three quality levels on the basis of the measurable firmness.

In addition to the firmness, which can be used as the main criterion for freshness and shelf life, further parameters were determined. First, the dry matter was evaluated. It should be noted that the correlation coefficient between strength and dry matter has a value of 0.77. In a multi-product calibration of 40 salad and 40 cherry tomatoes a PLSR model for the prediction of dry matter could be created. The RMSEP of the dry matter was 0.50 and 0.52, respectively, with the values of both varieties ranging between 5 % and 9 % dry matter.

To predict the brix value, a multi-product calibration for salad and cocktail tomatoes using PLSR was also made. The brix value of the tomatoes could be determined with an RMSEP of 0.56 and 0.68 °brix for salad and cocktail tomatoes in the range between 4 and 9 °brix .

3.3 Prediction of shelf life

In this regard, firmness can be used as one of the parameters to estimating shelf life of tomatoes. To this end, the time dependency over storage can be modeled by using the Arrhenius equations [14,15].

Data analysis is performed through the packet R software. The loss of firmness for all three-temperature levels as well as the final models for the three temperature levels are plotted in Figure 1.3. The kinetic parameters as well as the activation energy can then be extracted from the model. In Figure 1.3 the quality classes A, B and C are presented based on bonitur classifications. It shows that along the storage of tomatoes, the subjectively perceived firmness declines due to vapor pressure deficit and storage time. Since in this experiment, the vapore

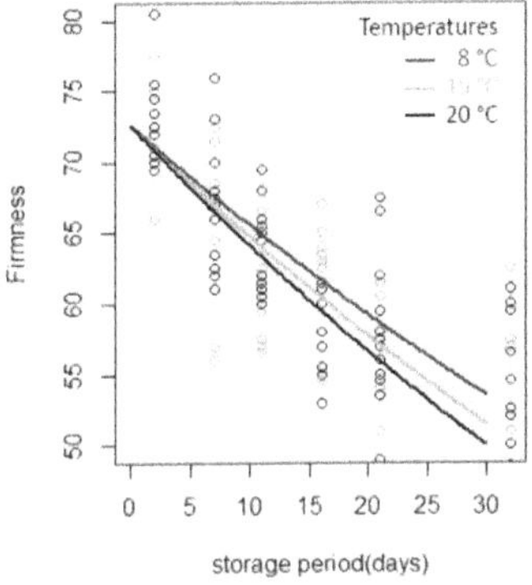

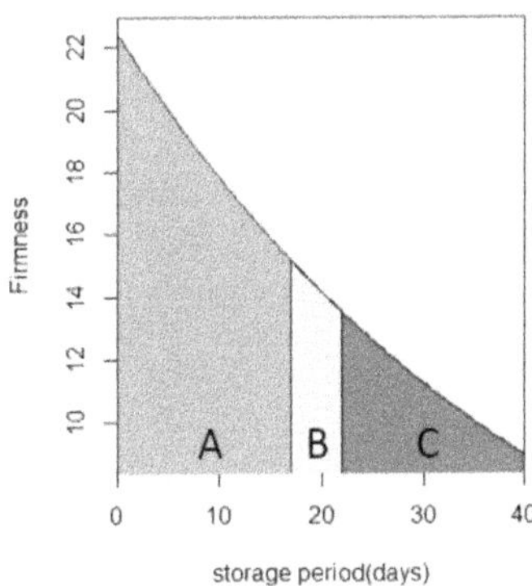

Figure 1.3: Firmness measured at 3 different temperature levels [8°,15°,20°C] and the fitted curves of the Arrhenius models (left). Firmness – storage time model for temperature 20°C indicating 3 different quality classes for salad tomatoes (right)

pressure deficit was relatively small, the influence of storage period is rated higher than the influence due to temperature. By considering the storage conditions at the point of sale with notably higher vapor pressure deficits (e. g. 50,2 hPa at 20°C and 50 % relative humidity) a faster decline in firmness can be expected.

4 Summary

Firmness is a strong parameter for quality of tomatoes. However, the measurement of firmness using classical measuring methods like penetrometers leads to fruit damage, manifesting itself in bruises and subsequently in a fast decay of the fruits. A similar practice can bee seen in supermarkets when consumers try to assess quality by touching the fruits. In both cases a non-destructive and non-contact optical measurement offers an advantage [16]. The results of this study show that firmness can be predicted with good accuracy using a miniaturized NIR-sensor. Furthermore, a relation between bonitur classifications and objective firmness measurements was established, indicating the possibility of distinguishing bonitur grades by means of NIR-spectroscopy. It is also shown that the decline of firmness over storage time with respect to vapor pressure deficit of tomatoes can be modelled. In combination, the decline of firmness as an indicator for shelf-life can be predicted using this portable NIR-Sensor.

Future work should address the validity of the PLSR correlation by adding new tomato varietes to the data set. Regarding the utilization of portable NIR-sensors by end-consumers these data sets should focus on varieties available in supermarkets.

Acknowledgement

This work was supported by the Bavarian Ministry of Food, Agriculture and Forestry as part of the alliance "Wir retten Lebensmittel".

References

1. UNECE, "Fresh fruit and vegetables - standards," 2017. [Online]. Available: https://www.unece.org/trade/agr/standard/fresh/ffv-standardse.html
2. A. Batu, "Determination of acceptable firmness and colour values of tomatoes," *Journal of Food Engineering*, vol. 61, no. 3, pp. 471–475, 2004.
3. M. Koch, "Fläche im unterglasanbau steigt weiter," 2018. [Online]. Available: https://www.ami-informiert.de/ami-maerkte/maerkte/ami-gartenbau/ami-meldungen-gartenbau/single-ansicht/singleview/news/artikel/flaeche-im-unterglasanbau-steigt-weiter.html
4. B. Rogge, "Blumenkohl mit kräftigem mengenplus in den top-10," 2018. [Online]. Available: https://www.ami-informiert.de/ami-maerkte/maerkte/ami-gartenbau/ami-meldungen-gartenbau/single-ansicht/singleview/news/artikel/blumenkohl-mit-kraeftigem-mengenplus-in-den-top-10.html
5. K. Flores, M.-T. Sánchez, D. Pérez-Marín, J.-E. Guerrero, and A. Garrido-Varo, "Feasibility in nirs instruments for predicting internal quality in intact tomato," *Journal of Food Engineering*, vol. 91, no. 2, pp. 311–318, 2009.
6. G. Kim, D.-Y. Kim, G. H. Kim, and B.-K. Cho, "Applications of discrete wavelet analysis for predicting internal quality of cherry tomatoes using vis/nir spectroscopy," *Journal of Biosystems Engineering*, vol. 38, no. 1, pp. 48–54, 2013.

7. C. A. T. dos Santos, M. Lopo, R. N. M. J. Páscoa, and J. A. Lopes, "A review on the applications of portable near-infrared spectrometers in the agro-food industry," *Applied spectroscopy*, vol. 67, no. 11, pp. 1215–1233, 2013.
8. M. Kranert, G. Hafner, J. Barabosz, H. Schuller, D. Leverenz, A. Kölbig, F. Schneider, S. Lebersorger, and S. Scherhaufer, "Ermittlung der weggeworfenen lebensmittelmengen und vorschläge zur verminderung der wegwerfrate bei lebensmitteln in deutschland," 2012.
9. Kusumiyati, T. Akinaga, M. Tanaka, and S. Kawasaki, "On-tree and after-harvesting evaluation of firmness, color and lycopene content of tomato fruit using portable nir spectroscopy," *Journal of Food, Agriculture & Environment*, vol. 6, no. 2, 2008.
10. G. Rateni, P. Dario, and F. Cavallo, "Smartphone-based food diagnostic technologies: A review," *Sensors (Basel, Switzerland)*, vol. 17, no. 6, 2017.
11. H. Kaur, R. Künnemeyer, and A. McGlone, "Comparison of hand-held near infrared spectrophotometers for fruit dry matter assessment," *Journal of Near Infrared Spectroscopy*, vol. 25, no. 4, pp. 267–277, 2017.
12. K. Ncama, L. S. Magwaza, C. A. Poblete-Echeverría, H. H. Nieuwoudt, S. Z. Tesfay, and A. Mditshwa, "On-tree indexing of 'hass' avocado fruit by non-destructive assessment of pulp dry matter and oil content," *Biosystems Engineering*, vol. 174, pp. 41–49, 2018.
13. F. Pedregosa, G. Varoquaux, A. Gramfort, V. Michel, B. Thirion, O. Grisel, M. Blondel, P. Prettenhofer, R. Weiss, V. Dubourg, J. Vanderplas, A. Passos, D. Cournapeau, M. Brucher, M. Perrot, and E. Duchesnay, "Scikit-learn: Machine learning in Python," *Journal of Machine Learning Research*, vol. 12, pp. 2825–2830, 2011.
14. J. Pinheiro, C. Alegria, M. Abreu, E. M. Gonçalves, and C. L. Silva, "Kinetics of changes in the physical quality parameters of fresh tomato fruits (Solanum lycopersicum, cv. 'Zinac') during storage," *Journal of Food Engineering*, vol. 114, no. 3, pp. 338–345, feb 2013. [Online]. Available: https://www.sciencedirect.com/science/article/abs/pii/S0260877412004074
15. C. Van Dijk, C. Boeriu, F. Peter, T. Stolle-Smits, and L. Tijskens, "The firmness of stored tomatoes (cv. Tradiro). 1. Kinetic and near infrared models to describe firmness and moisture loss," *Journal of Food Engineering*, vol. 77, no. 3, pp. 575–584, dec 2006. [Online]. Available: https://www.sciencedirect.com/science/article/abs/pii/S0260877405005224
16. M. Geyer, M. Linke, I. Gerbert, O. Schlüter, and H.-P. Kläring, "Beurteilen der Haltbarkeit klimakterischer Früchte am Beispiel der

Tomate," *LANDTECHNIK – Agricultural Engineering,* vol. 63, no. 3, pp. 158–159, jun 2008. [Online]. Available: https://www.landtechnik-online.eu/ojs-2.4.5/index.php/landtechnik/article/view/2008-3-158-159

7 Comparison of colorimeter and different portable food-scanners for non-destructive prediction of lycopene content in tomato fruit

In: Postharvest Biology and Technology 167 (111232), p. 1-8

Cite as: Goisser, S.; Wittmann, S.; Fernandes, M.; Mempel, H.; Ulrichs, C. (2020): Comparison of colorimeter and different portable food-scanners for non-destructive prediction of lycopene content in tomato fruit. Postharvest Biology and Technology 167 (111232), p. 1-8.

DOI: https://doi.org/10.1016/j.postharvbio.2020.111232

Title

Comparison of colorimeter and different portable food-scanners for non-destructive prediction of lycopene content in tomato fruit

Author names and affiliations

Simon Goisser[a,c], Sabine Wittmann[a], Michael Fernandes[b], Heike Mempel[a], Christian Ulrichs[c]

[a]University of Applied Sciences Weihenstephan-Triesdorf, Greenhouse Technology and Quality Management, Am Staudengarten 10, 85354 Freising, Germany

[b]Deggendorf Institute of Technology, Technology Campus Grafenau, Department of Applied Artificial Intelligence, Hauptstraße 3, 94481 Grafenau, Germany

[c]Humboldt Universität zu Berlin, Faculty of Life Sciences, Division Urban Plant Ecophysiology, Lentzeallee 55/57, 14195 Berlin, Germany

simon.goisser@hswt.de, sabine.wittmann@hswt.de, michael.fernandes@th-deg.de, heike.mempel@hswt.de, christian.ulrichs@hu-berlin.de

Corresponding author

Simon Goisser, Am Staudengarten 10, 85354 Freising, Germany (simon.goisser@hswt.de)

Abstract

Lycopene, the red colored carotenoid in tomatoes, has various health benefits for humans due to its capability of scavenging free radicals. Traditionally, the quantification of lycopene requires an elaborate extraction process combined with HPLC analysis within the laboratory. Recent studies focused simpler methods for determining lycopene and utilized spectroscopic measurement methods. The aim of this study was to compare non-destructive methods for the prediction of lycopene by using color values from colorimeter measurements and VIS/NIR spectra recorded with three commercially available and portable VIS/NIR spectrometers, so called food-scanners. Tomatoes of five different ripening stages (green to red) as well as tomatoes stored up to 22 days after harvest were used for modeling. After measurement of color values and collection of VIS/NIR spectra the corresponding lycopene content was analyzed spectrophotometrically. Applying exponential regression models yielded very good prediction of lycopene for color values L*, a*, a*/b* and the tomato color index of 0.94, 0.90, 0.90 and 0.91 respectively. Color value b* was not a suitable predictor for lycopene content, whereas the $(a^*/b^*)^2$ value had the best linear fit of 0.87. In comparison to color measurements, the cross-validated prediction models developed for all three food-scanners had coefficients of determination (r^2_{CV}) ranging from 0.92 to 0.96. Food-scanners also can be used for additional measurements of internal fruit quality, and therefore have great potential for fruit quality assessment by measuring a multitude of important fruit traits in one single scan.

Keywords

Tomato; *Solanum lycopersicum L.*; Colorimeter; Non-destructive; Portable near-infrared spectrometer; Lycopene content; Food-scanner

1. Introduction

Tomato is one of the most important vegetable crops cultivated in Germany, with a total of 398 ha and therefore a percentage of 30 % of Germany's greenhouse production area for vegetables (AMI 2019). Due to the high competition with imported produce, fruit quality is of particular importance. Several internal quality parameters such as sugar content, acidity and brix/acid ratio help to describe the sweetness and taste of tomatoes and constitute an indicator for fruit maturity due to changes of these biochemical compounds during ripening (Gautier et al. 2008). Additional important indicators of tomato quality, which determine shelf life and influence consumer acceptance, are fruit firmness and color (Batu 2004). The distinctive skin color of red tomato cultivars results from the carotenoid and precursor of β-carotene and vitamin A, lycopene, (Cazzonelli 2011). Lycopene and other carotenoids support plants in their photoprotection against excess light by scavenging free radicals such as singlet molecular oxygen (Sies und Stahl 1995), where lycopene in particular is known to be the carotenoid with the most efficient biological singlet oxygen quenching capabilities (Di Mascio et al. 1989). In addition to the importance of lycopene on the metabolism of plants, the substance is also believed to have an anti-carcinogenic effect in humans (Stahl et al. 2001; Seren et al. 2008). Some breeding companies are making use of this supposed health benefit and advertise lycopene enriched tomato cultivars such as 'Lyterno', 'Lynato' or 'Lycotom' (Kugler 2009; Rijk Zwaan 2019). However, use of these varieties only adds value if lycopene concentrations can be measured and verified using simple methods.

The quantification of components like lycopene can be extremely elaborate, requiring wet chemical extraction procedures of tomato products and subsequent HPLC and/or spectrophotometric measurements (Sadler et al. 1990). Lycopene content of tomato at different ripening stages can be measured using L*, a*, b* color readings as well as a*/b* and $(a^*/b^*)^2$ color ratios (Arias et al. 2000; Brandt et al. 2006). Clément et al. (2008a) also used L*, a* and b* values for the calculation of the tomato color index (TCI). Lycopene content of tomatoes and tomato products was assessed using VIS/NIR spectroscopy (Clément et al. 2008b; Li et al. 2017; Tilahun et al. 2018). The performance of portable NIR instruments for determining lycopene content of tomato fruit on-vine in the field has been tested (Kusumiyati et al. 2008).

Decreasing price and size of portable NIR devices has led to development of compact NIR spectrometers, e.g. the F-750 Produce Quality Meter (Felix Instruments 2017) specifically designed for in-field evaluation of fruit quality. In addition, miniaturized hand-held NIR-devices called food-scanners have been developed, often designed for end-consumers to work wirelessly in combination with a tablet or smartphone and operate on a cloud-based platform. They aim to provide relevant information on food quality traits such as protein, sugar or total energy content, calories, allergens, and contaminants (Rateni et al. 2017). Recent studies on horticultural and agricultural products demonstrate good results using a SCiO™ molecular Sensor (Consumer Physics) for the prediction of dry matter in apples, kiwifruit and stone fruit (Kaur et al. 2017), total soluble solids in kiwifruit (Li et al. 2018), total antioxidant capacity in gluten-free grains (Wiedemair und Huck 2018), and prediction of tomato shelf life (Goisser et al. 2019a). Sheng et al. (2019) used a self-developed miniaturized NIR spectrometer and corresponding mobile application for the determination of lycopene content in tomato fruit. In conclusion, miniaturized NIR devices seem to be a suitable method of choice in measuring various quality parameters of fresh produce in a fast and non-destructive way.

The objective of this study was to develop prediction models for lycopene content of tomato fruit using three commercially available portable food-scanners. Partial least squares (PLS) prediction models obtained from these food-scanners are compared to regression models for lycopene derived from color values through colorimeter measurements. Additionally, this study investigates the practicability of food-scanner applications along the horticultural supply chain by analyzing the lycopene content of tomatoes of different ripening stages and applying existing pre-processing methods of spectra as used by food-scanner manufacturers.

2. Materials and methods

2.1 Sample material

Tomatoes (*Solanum lycopersicum* L., cv. Avalantino) from a greenhouse of the University of Applied Sciences Weihenstephan-Triesdorf were used for this study. 'Avalantino' is a red-colored, smooth-surfaced and round-shaped salad cultivar commercially grown in Germany yielding fruit with two locules, weighing between 70 - 85 g and with diameter of 55 – 60 mm. Plants were grown during summer 2019 under commercial growth conditions using a hydroponic recirculating drip irrigation system. During cultivation and harvesting of fruit for this experiment, the average air temperature and relative humidity within the greenhouse were 20.1 ± 8.1 °C and 74.5 ± 17.7 %, respectively.

2.2 Experiment design

The experiment was divided into assessments of ripening stage and ripening during storage (Fig. 1). For the first, fruit of different ripening stages were harvested from middle of July until start of August 2019. The USDA Color Chart (USDA 1975) was used as orientation and five different ripening stages were determined for harvest: 1) Green (tomato surface completely green in color) 2) Breaker (break in color from green to yellow) 3) Turning (change in color from yellow to orange) 4) Light red (shift in color from orange to light red) 5) Red (tomato surface completely covered in full red color). Measurements were carried out from July 18 to the first day of August on 5 days at two to five day intervals on continuous and fixed time intervals (07:00 a.m. to 02:00 p.m.). Three fruits of each ripening stage were randomly selected and harvested daily, resulting in a batch of 15 fruit each day and a total of 75 fruit. Fruit were transported to University facilities where the calyx was removed, and numbered prior to acquisition of spectra and consecutive reference measurements.

For the second part, 90 fruit of the ripening stage 5 were harvested in the fourth week of August 2019 and transported to University facilities. Afterwards, the calyx was removed, fruit were numbered and placed on a metal grid to allow full air circulation at room temperature of 22.4 ± 1.7 °C and relative humidity of 67.4 ± 3.0 % for the storage period of 22 days. Measurements were performed in batches of 15 randomly selected fruit on the day of harvest and 8, 13, 17, 20, and 22 days after harvest (DAH). All spectral and corresponding reference measurements were conducted on a single fruit basis.

Sample composition

Experiment	Section 1	Section 2
Sample pool	Greenhouse fruit	Stored fruit
Sample selection	Random	
Sample number	15 fruit at 5 stages	15 fruit
Number of repetitions	5	6
Total sample number	75	90
Sample preparation	Calyx removal, Numbering of fruit	

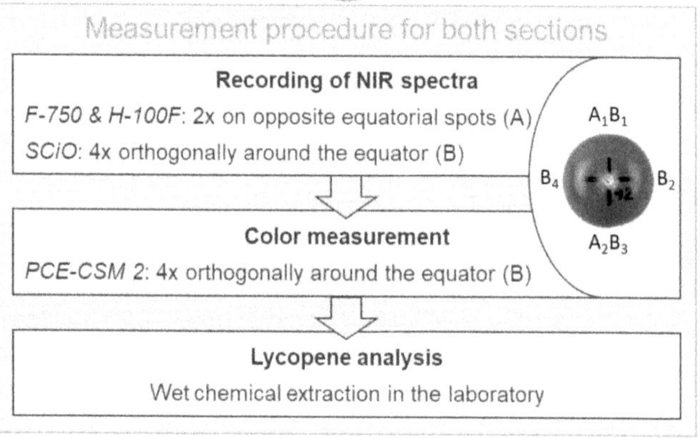

Fig. 1. Sample composition and measurement procedure for both sections of the experiment

2.3 Portable NIR instruments

Three commercially available portable NIR spectrometers were used:; the F-750 Produce Quality Meter (Firmware v.1.2.0 build 7041, Felix Instruments, Portland, USA); SCiO™ (version 1.2, Consumer Physics, Hod HaSharon, Israel) and the H-100F (Sunforest, Incheon, Korea). The characteristics and operating parameters of these devices are summarized in Table 1.

2.4 Acquisition of spectra

The measurement method for all three devices is diffuse reflection. For the F-750 and H-100F, two spectra for each fruit were taken on the same location on opposite sides at the equator.

The F-750 uses a shutter with gold-coated foil as its reference standard, which is calculated for every sample measurement, therefore an additional white reference was not needed. Spectra were recorded using the built-in training set method. Recording of spectra with the H-100F was and referenced against the white standard attached in the protective plastic cap of the measurement head at the beginning and after every 20 measurements. Spectra were recorded by linking the device via USB to a computer, using the Sunforest H-100 Laboratory Program (V 1.1, Sunforest, Incheon, Korea) for displaying and saving of raw spectra. Spectra for the SCiO™ were recorded on four orthogonally and equidistant points around the equator of each fruit, including the two measurement spots used with the F-750 and H-100F. Since the LED light source of the SCiO™ illuminates a distinct smaller surface area compared with the halogen lamps of the F-750 and H-100F, four measurement points instead of two were deemed more suitable to depict an appropriate average fruit spectra. Analogous to the H-100F, the white standard within the SCiO™ protective case was used for referencing at the beginning and after every 20 performed measurements. To be able to record raw spectra with the SCiO™, a development license is needed in combination with a smartphone application. During recording, spectra were transferred from the device via Bluetooth to the smartphone, where internet connection was used to store the spectral information in a provided cloud-service.

2.5 Reference measurement of fruit quality parameters

Measurement of color values and lycopene content followed a fixed operating procedure and was performed immediately after recording of spectra. Color was measured with a colorimeter (PCE-CSM 2, PCE Instruments, Meschede, Germany) on four spots orthogonally around the fruit equator, matching the spots of NIR spectra acquisition. The readings were averaged by the colorimeter and used for further reference. Averaged values were used for computing a^*/b^* and $(a^*/b^*)^2$ ratios as well as the tomato color index [$TCI = 2000a^*/L^*(a^{*2}+b^{*2})^{1/2}$] provided by Clément et al. (2008a) for each fruit.

Complete fruit were subsequently blended using a conventional smoothie mixer (WMF Kult X Mix&Go 300 Watt, WMF Group GmbH, Geislingen/Steige, Germany). 3-4 g of each purée was transferred to a small glass beaker and diluted with deionized water. Beakers were sealed with Parafilm® and placed in an iced-cooled and light-tight polystyrene box for transportation to the laboratory for lycopene analysis.

The method of Fish et al. (2002) was used with some modification for determination of lycopene content of duplicate diluted purée samples. A solvent solution of 10 mL ethanol, 10 mL acetone with 0.05% (w/v) butylated hydroxytoluene (BHT) and 20 mL n-hexane was prepared in 100 mL amber vials. 1 ml of stirred tomato purée was added to the vial on a digital laboratory scale using an automatic pipette. Vials were put on their sides in a polystyrene box, covered with ice packs and the box placed on an orbital shaker (type 3015, GFL mbH, Burgwedel, Germany) to mix at 150 rpm for 15 min. After shaking, 3 mL of deionized water was added to each vial and samples shaken by hand. Vials were then left at room temperature for 5 min to allow for phase separation. The absorbance of the upper hexane layer was measured at 503 nm versus a blank of hexane solvent with a spectrophotometer (Specord 200 Plus, Analytik Jena AG, Jena, Germany). The lycopene content of each fruit was expressed as mg kg^{-1} by using the equation provided by Fish et al. (2002), accounting for weights of diluted purée and ratio of purée to deionized water prior to preparation of diluted purée.

2.6 Statistical and multivariate analysis

Data analysis and determination of statistical significance between ripening stages of the fruit was performed with Minitab software (version 18.1, Minitab Inc., Munich, Germany) using one-way ANOVA and Tukey's test ($p = 0.05$). Microsoft® Excel® 2016 was used for the calculation of Pearson's correlation coefficients, linear and exponential regressions as well as the determination of equations for the best calculation of lycopene with respect to each color value.

The evaluation of a linear relationship between food-scanner spectra and lycopene content was performed with The Unscrambler® software (version 10.5.1, CAMO, Oslo, Norway). The software offers various options of data pre-processing and subsequent evaluation. Whereas pre-processing techniques were selected based on the best result for model calibration in other studies (Li et al., 2018), the pre-processing settings provided by the respective manufacturer's software were taken into account to evaluate the practicability of existing standard settings.

The SCiO™ device, has a cloud-based browser application for building of prediction models from raw spectra and respective reference data. Raw spectra were downloaded from the cloud-based samples library as an Excel-file and analyzed with The Unscrambler®. The following pre-processing settings were applied: Logarithm, averaging all four scans per sample, first derivative (window of 35 and polynom degree of 2), selecting the full wavelength range from 740 - 1070 nm and subtracting the average. The algorithm was set to partial least square regression (PLSR) and 20 random segments were used for cross-validation.
The F-750 device uses a free software tool for analysis of spectra and building of prediction models. Spectra were loaded into the current Model Builder Software (version 1.3.0.192 BETA, Felix Instruments, Portland, USA). Spectra in second derivative form in the wavelength range from 477 - 1059 nm was selected, transferred to an Excel-file and subsequently loaded into The Unscrambler® for evaluation. Besides averaging of both spectra per sample no further

pre-processing was applied and PLSR algorithm used for building of prediction models, using 20 random segments for cross-validation.

Spectra from the H-100F were saved after recording from the Sunforest H-100 Laboratory Program and then analyzed with The Unscrambler® software. Spectra were provided in second derivative form in the range from 650 - 950 nm and used without further pre-processing after averaging the two scans per sample. PLSR algorithm was used for model building. Cross-validation was performed by using 20 random segments.

3. Results

3.1 Change of color values and lycopene content across ripening and storage

The average lycopene content, corresponding color readings as well as the derived a*/b*, (a*/b*)^2 and TCI values of each ripening stage including standard deviations and significant differences are shown in Table 2. The lowest lycopene content, 0.84 mg kg^{-1}, was measured in green fruit. Contents increased with ripening stage to reach 118.6 mg kg^{-1} for red fruit and accumulate after harvest to 153.9 mg kg^{-1} in stored fruit on day 13, and then decreased to 133.7 mg kg^{-1}.

3.2 Correlation of color readings and lycopene content

The color readings produced with the colorimeter as well as the derived a*/b*, (a*/b*)^2, TCI values showed different relations to the lycopene contents. All values, with the exception of b*, have high linear correlations to the lycopene content. In addition to the Pearson correlations, both linear and exponential regressions where evaluated to determine the regression with the best fit (Table 3). Exponential regression of L* and lycopene content produced a higher R^2 (0.94) than the linear regression. Whereas lycopene content increased during ripening, fruit color changed from light to dark, expressed as a decrease of L*. Fig. 2A illustrates this exponential decay of L* during lycopene accumulation.

Color value a* displays the shift from green (-) to red (+) and is therefore directly linked to the accumulation of lycopene in fruit. As depicted in Fig. 2B, this relation can be best described as an exponential rise (R^2 = 0.90) using an exponential function (Table 3). Neither linear nor exponential regression produced a good fit for b* and lycopene content. Therefore, b* is not suitable for the prediction of lycopene content in the fruit (Table 3).

As shown in Table 3, the exponential regression of the a*/b* ratio produced a slightly better fit (R^2 = 0.90) compared with linear regression (R^2 = 0.88), whereas the (a*/b*)^2 value showed a superior fit for the linear regression (Fig. 2C). Similar to a* and the a*/b* ratio, the derived TCI values and lycopene produced a good fit (R^2 = 0.91) applying an exponential regression (Fig. 2D).

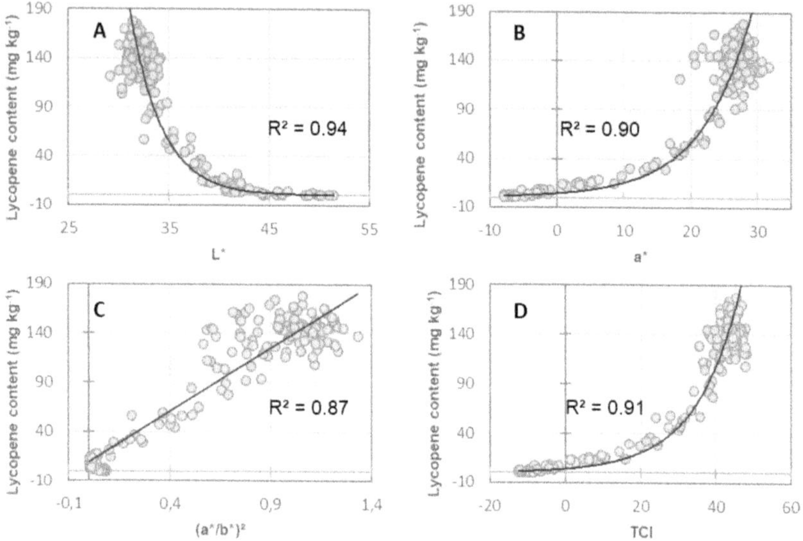

Fig. 2. Regressions showing the best fit between lycopene content and L* (A), a* (B), $(a^*/b^*)^2$ ratio (C) and TCI (D)

3.3 Correlation of food-scanner spectra and lycopene content

The VIS/NIR spectra of intact tomato fruit recorded with all three food-scanners were used to develop PLS models for calibration and cross validation (Table 4).

After cross validation the models for the prediction of lycopene showed high coefficients of determination in cross validation r^2_{CV} (> 0.90) for all three devices. Model of best performance with respect to high r^2_{CV} and low $RMSE_{CV}$ was obtained for the F-750 device (r^2_{CV} = 0.96, $RMSE_{CV}$ = 12.3 mg kg^{-1}) closely followed by the H-100F (r^2_{CV} = 0.95, $RMSE_{CV}$ = 14.3 mg kg^{-1}) and SCiO™ instrument (r^2_{CV} = 0.92, $RMSE_{CV}$ = 17.1 mg kg^{-1}) as illustrated in Fig. 3.

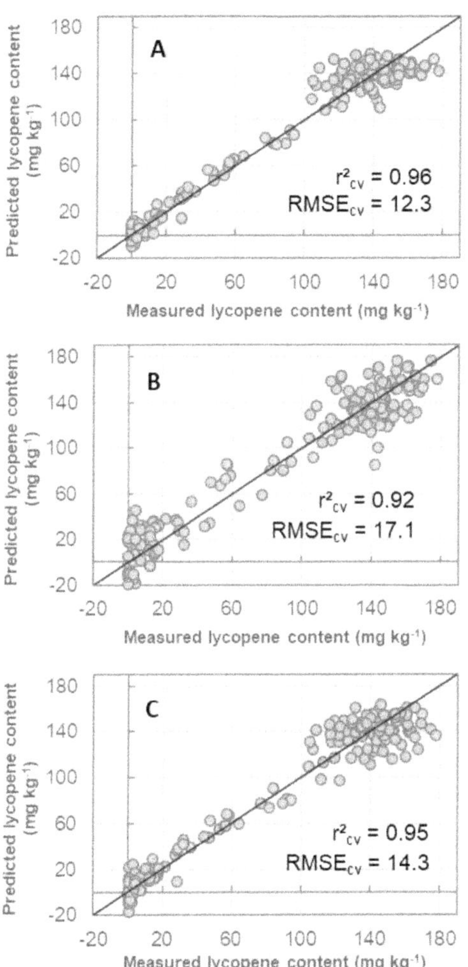

Fig. 3. Correlation between measured and predicted lycopene content for cross validated models of the F-750 Produce Quality Meter (A), SCiO™ Molecular Sensor (B) and H-100F (C)

4. Discussion

4.1 Prediction of lycopene using color readings

The results of this study illustrate the usability of colorimeters and food scanners for the non-destructive determination of lycopene in tomato fruit. Various studies investigated the possibility of estimating the lycopene concentration of tomato fruit using only color readings via colorimeter. Applying linear regression for lightness L* yielded R^2 = 0.79 in this study, which is in the range of findings from previous work (R^2 = 0.64 to 0.82) using three tomato cultivars at green to red-ripe stage (D'Souza et al. 1992). Results for linear regressions of L* value on whole tomato fruit and lycopene calculated in similar studies showed poorer performance, ranging from R^2 = 0.29 (Hyman et al. 2004), R^2 = 0.50 (Tilahun et al. 2018) to R^2 = 0.67 (Carvalho et al. 2005). Nevertheless, linear regression performed by Arias et al. (2000) showed better fit for L* (R^2 = 0.84) compared with the present study. They also found that the fit of an exponential regression was superior to that of linear regression, with similar results (R^2 = 0.93) to those in our study (R^2 = 0.94).

Utilizing linear regression for the determination of lycopene content and a* resulted in a good fit (R^2 = 0.85), which is superior to findings achieved on intact tomatoes reported in the literature reaching from R^2 = 0.31 (Hyman et al. 2004), R^2 = 0.55 - 0.68 (D'Souza et al. 1992) to R^2 = 0.79 (Carvalho et al. 2005). However, a recent study demonstrated even better correlation of R^2 = 0.90 (Tilahun et al. 2018). Applying exponential regression compared to linear regression obtained notably better fit of R^2 = 0.96 compared to R^2 = 0.80 (Arias et al. 2000), which is in line with findings from our study.

Similar to various previous research (D'Souza et al. 1992; Carvalho et al. 2005; Hyman et al. 2004; Arias et al. 2000), the results in our study for color value b* indicate no suitable predictive capabilities with respect to lycopene content in the fruit. In contrast, Tilahun et al. (2018) reported a coefficient of determination for lycopene prediction using color value b* (R^2 = 0.71). However, the authors analyzed only three ripening stages (breaker, pink, and red) of a cultivar, which had lower lycopene contents and readings in color values than cultivar that we used. Different ranges of reference values could influence the linearity of the color value b*.

Prediction models for lycopene content using linear regression of a*/b* values also obtained good results in this study (R^2 = 0.88), which is identical to findings reported by Arias et al. (2000) and Carvalho et al. (2005) and superior to results obtained in other studies (D'Souza et al. 1992; Hyman et al. 2004). An even better coefficient of determination (R^2 = 0.98) was reported by Tilahun et al. (2018). The utilization of an exponential regression for the correlation of a*/b* with lycopene in the current study resulted in a better fit (R^2 = 0.90) compared to linear regression; consistent also with findings from Arias et al. (2000) and Brandt et al. (2006), who achieved an even superior exponential regression fit of R^2 = 0.96 and 0.92 respectively compared to our result.

Various previous studies applying simple linear regression identified the best fit for lycopene prediction models by using the $(a^*/b^*)^2$ value, ranging from R^2 = 0.71 - 0.83 (D'Souza et al. 1992), R^2 = 0.73 (Hyman et al. 2004), R^2 = 0.89 (Carvalho et al. 2005) to R^2 = 0.90 (Arias et al. 2000). In contrast to these reports, the present study achieved slightly better linear regressions by using a*/b* as well as TCI (R^2 = 0.88 in either case) compared to $(a^*/b^*)^2$ (R^2 = 0.87). In accordance with findings reported by Arias et al. (2000), the $(a^*/b^*)^2$ value seems to be the only color variable best suitable for linear regression, whereas L*, a* and a*/b* showed the best fit for exponential regression.

The TCI has been used as indicator of tomato ripening (Chen et al. 2008; Clément et al. 2008b). Clément et al. (2008a) obtained a linear correlation coefficient between TCI and lycopene content of 0.83. The Pearson's correlation coefficient in the present study resulted in a higher correlation of 0.94, which seems to be best described using an exponential compared to a linear regression (Table 3).

4.2 Prediction of lycopene using food-scanner spectra

Vis/NIR spectroscopy has been utilized in several studies to examine the prediction of lycopene content in tomato products. The application of a benchtop NIR spectrometer for the prediction of lycopene content in the fruit and various tomato products (ketchup, purée, mash) yielded a prediction model of r^2 = 0.85 and standard error of cross-validation (SE_{CV}) = 9.12 mg kg^{-1} (Baranska et al. 2006). In contrast to the present study, tomatoes were mashed prior to the collection of spectral data. Szuvandzsiev et al. (2014) analyzed 16 tomato cultivars with a portable spectrometer by acquiring spectra and measuring lycopene content of purée and obtained a prediction model of r^2 = 0.75 and $RMSE_{CV}$ = 19.9 mg kg^{-1}. Other studies investigated the predictability of lycopene content by scanning fruit using reflectance spectra from benchtop NIR instruments, resulting in prediction models of r^2 = 0.91 and $RMSE_{CV}$ = 8.76 (Li et al. 2017) and r^2 = 0.98, $RMSE_{CV}$ = 3.15 (Clément et al. 2008b). The utilization of a portable NIR instrument for the evaluation of lycopene content of fruit on-vine yielded prediction models of r^2 = 0.81 and SEP = 82,7 mg kg^{-1} (Kusumiyati et al. 2008). Analysis of transmittance instead of reflectance spectra within a laboratory measurement setup of fruit resulted in PLS prediction models of r^2 = 0.85 and SEP = 1.79 mg kg^{-1} (Tilahun et al. 2018). Sheng et al. (2019) used a commercially available miniaturized NIR spectrometer (DLP® NIR scan Nano, Texas Instruments) within a self-developed device and corresponding mobile application and achieved a prediction of r^2 = 0.80 and RMSEP = 7.45 mg kg^{-1} for lycopene content in fruit. In contrast to these previous studies, we adopted a different approach by using intact tomato fruit. In addition, portable and miniaturized devices were used in our study instead of laboratory NIR instruments. By measuring fruit across all ripening stages with three commercially available portable food-scanners and applying standard settings of the manufacturer's software, this study focused on the practicability of existing mobile NIR devices.

Correlation coefficients obtained in this study for the three portable NIR devices often surpassed those reported in the literature (r^2 > 0.90). Previous research has used ripe fruit bought from the supermarket (Clément et al. 2008b; Sheng et al. 2019) or reduced the scope of the experimental setup to selected ripening stages (breaker, pink, and red) within the cultivation process (Tilahun et al. 2018). This study demonstrated the importance of obtaining tomatoes with high variations of lycopene content by using fruit across all ripening stages during cultivation (green, breaker, turning, pink, red) and subsequent storage to obtain prediction models with high predictive capabilities.

Compared to previous studies, $RMSE_{CV}$ values exceeded those reported in the literature. Considering additional correction and pre-processing techniques of spectra such as methods for temperature change (Sheng et al. 2019) as well as optimization of PLS algorithms (Li et al. 2017) could lead to a further improvement of model performance.

The comparison of the three food-scanners used in this study shows only slight differences with respect to the predictive capability of lycopene content of fruit. The spectrometer of the F-750 utilizes the bandwidth of 477 - 1059 nm, covering the visible range and therefore the shift in color values such as green and red, which is closely correlated to the synthesis of lycopene. Consequently, a prediction model with high accuracy for the prediction of lycopene content could be modeled. Although operating in a smaller spectral range (650 - 950 nm), the H-100F still covers the visible red spectrum and therefore achieves a model performance only slightly inferior to the F-750. The SCiO™ Molecular Sensor applies the wavelength range from 740 - 1070 nm, hence only a small range of the visible red spectrum which covers lycopene synthesis can be recorded and used for model building. Consequently, the prediction model of the SCiO™ shows a slightly lower accuracy compared to the F-750 and H-100F. Nevertheless, all three food-scanners achieved predictions of high accuracy ($r^2 > 0.90$) for lycopene content by applying pre-processing methods of spectra which come standard with these devices. To evaluate the practicability of three existing food-scanner systems, this study focused on the utilization of these standard settings and used fruit of all ripening stages. The results of this study emphasize the informative value of these small mobile devices and indicate the great predictive potential of NIR food-scanners. Further optimization of spectral pre-processing could improve these existing standard settings and yield better results for model calibration as Li et al. (2018) demonstrated for the SCiO™ device.

4.3 Colorimeter vs. food-scanner

By drawing a comparison between the accuracy (R^2) of color values best suitable for the prediction of lycopene (L*, a*, a*/b*, TCI) and coefficient of determination (r^2) for models developed using food-scanners, only slight differences can be identified. The evaluation of both types of measurement instruments revealed prediction models of high accuracy (R^2 and $r^2 > 0.90$). Therefore, both methods for the prediction of lycopene content of fruit are deemed feasible. In general, the non-destructive determination of lycopene from fruit by using colorimeter and food-scanner allows a fast, cheap and timely prediction compared with wet chemical extraction processes in the laboratory. Whereas the application of a colorimeter allows the measurement of chromaticity values and consequently the lycopene content of tomatoes by using previously established equations, VIS/NIR spectrometers can be used as multidimensional tools for the prediction of various quality traits derived from a single scan. Elaborate, destructive and time-consuming measuring procedures (e.g., refractometric measurements, chemical extractions) of these quality traits could therefore be replaced by the use of food-scanners.

Results from this experiment indicate that existing and commercially available NIR instruments have the capability of measuring lycopene content of tomato fruit. In comparison to elaborate NIR measurements using benchtop devices the food-scanners used in this study feature a simple handling and measurement procedure. Additionally, the standard pre-processing methods for spectra that come with these devices yielded prediction models of high accuracy. Due to this practicability, future applications of food-scanners on various levels of the fruit and vegetable supply chain will become more feasible.

Previous studies indicate that NIR spectroscopy can be used to not only predict lycopene content, but also β-carotene concentration, of tomato fruit (Baranska et al. 2006; Tilahun et al. 2018). Moreover, portable and miniaturized NIR instruments have successfully been used for the determination of important tomato fruit quality parameters such as firmness and sugar content (Kusumiyati et al. 2008; Sheng et al. 2019; Goisser et al. 2019a). As illustrated by Kaur et al. (2017), the devices used in this study are also capable of predicting important quality traits like fruit dry matter in apple, kiwifruit and stone fruit with good to high accuracy. Other portable NIR instruments have been reported for the analysis of various quality attributes on a variety of produce (dos Santos et al. 2013). Therefore, food-scanners can be used on a multitude of fruit and vegetables for the non-destructive determination of various quality parameters simultaneously, provided that these parameters are quantifiable by NIR spectroscopy and corresponding prediction models are available. Novel approaches by means of using cloud-based databases in combination with advanced algorithms and vast material libraries (Rateni et al. 2017) indicate the direction in which these miniaturized instruments are heading. The results of a qualitative study conducted with actors of the German fresh produce supply chain indicate that food-scanners could facilitate future quality control on various supply chain levels of fruit and vegetables by providing a non-destructive, fast and objective measurement method (Goisser et al. 2019b).

5. Conclusion

Use of a colorimeter and three food-scanners for the predictability of lycopene content in tomato fruit has been compared and an overview of the various possibilities and accuracies for non-destructive lycopene measurement provided. The comparisons highlight the potential of these devices. Food-scanners provide the possibility in integrating several internal quality measurements within a single scan, therefore enabling users to gain a more thorough and objective picture of fruit quality. Food-scanners would replace elaborate, time-consuming and work-intensive destructive measurements of internal fruit quality. As indicated by the good performance of standard pre-processing methods of spectra, users have the opportunity to create their own models without in-depth knowledge in the field of NIR spectroscopy and model building, since manufacturers provide simple software solutions.

Our results are consistent with previous studies. Such studies assumed simple linear regression models for color values and lycopene content, but direct comparison between linear and exponential models indicated an advantage in choosing exponential functions for L*, a* and a*/b* values for the prediction of lycopene contents. The b* value was not suitable as a predictor of lycopene content, while the $(a^*/b^*)^2$ value yielded the best fit using a linear regression model. TCI as indicator of fruit ripeness used in different earlier studies was a good predictor of lycopene content using an exponential regression, a feature previously not described in the scientific literature.

The ability of a fast, easy and non-destructive measurement of lycopene leads to greater opportunities in promoting individual tomato cultivars with high lycopene contents in an easy and replicable manner along the supply chain. On the one hand, breeders and producers could use these devices to promote cultivars with high lycopene contents. On the other hand, definition and monitoring of specifications for lycopene contents are made possible by simple and fast measurements. Consequently, health-related quality traits could gain greater importance in marketing within a very price-sensitive segment. To develop more valid and robust prediction models for lycopene content, future research should focus on the selection of the best pre-processing techniques. By adding additional tomato cultivars with different concentrations of lycopene content to the current data set as well as using external samples for validation, the robustness of the current prediction model could be increased.

Acknowledgements

The research was supported by the Bavarian Ministry of Food, Agriculture and Forestry as part of the alliance "Wir retten Lebensmittel" [We Save Foodstuffs] as well as the QS Science Funds in Fruit, Vegetables and Potatoes. The authors thank STEP Systems GmbH from Nuremberg, Germany for their provision of two food-scanner devices during the course of this research project. The authors wish to acknowledge the help of Dr. Dieter Lohr during the experimental setup of the lycopene analysis as well as Prof. Dr. Eckhard Jakob and his laboratory technicians for providing a laboratory place.

References

AMI (2019): Mehr Gemüse unter Glas angebaut. Unter Mitarbeit von Sonja Illert. Hg. v. Agrarmarkt Informations-Gesellschaft mbH. Online verfügbar unter https://www.ami-informiert.de/ami-maerkte/maerkte/ami-gartenbau/ami-meldungen-gartenbau.

Arias, R.; Lee, T. C.; Logendra, L.; Janes, H. (2000): Correlation of lycopene measured by HPLC with the L, a, b color readings of a hydroponic tomato and the relationship of maturity with color and lycopene content. In: *Journal of agricultural and food chemistry* 48 (5), S. 1697–1702. DOI: 10.1021/jf990974e.

Baranska, M.; Schütze, W.; Schulz, H. (2006): Determination of lycopene and beta-carotene content in tomato fruits and related products: Comparison of FT-Raman, ATR-IR, and NIR spectroscopy. In: *Analytical chemistry* 78 (24), S. 8456–8461. DOI: 10.1021/ac061220j.

Batu, Ali (2004): Determination of acceptable firmness and colour values of tomatoes. In: *Journal of Food Engineering* 61 (3), S. 471–475. DOI: 10.1016/S0260-8774(03)00141-9.

Brandt, Sára; Pék, Zoltán; Barna, Éva; Lugasi, Andrea; Helyes, Lajos (2006): Lycopene content and colour of ripening tomatoes as affected by environmental conditions. In: *J. Sci. Food Agric.* 86 (4), S. 568–572. DOI: 10.1002/jsfa.2390.

Carvalho, Wesley; Fonseca, Maria Ester; Silva, Henoque R.; Boiteux, Leonardo S.; Giordano, Leonardo B. (2005): Estimativa indireta de teores de licopeno em frutos de genótipos de tomateiro via análise colorimétrica. In: *Horticultura Brasileira* 23 (3), S. 819–825. DOI: 10.1590/S0102-05362005000300026.

Cazzonelli, Christopher I. (2011): Carotenoids in nature: insights from plants and beyond. In: *Functional Plant Biology* 38 (11), S. 833–847. DOI: 10.1071/FP11192.

Chen, Limei; Raghavan, G. S. V.; Charlebois, Denis; Charles, Marie Therese; Vigneault, Clément (2008): Non-destructive measurement of tomato quality using visible and near-infrared reflectance spectroscopy. In: *Presentation at the CSBE/SCGAB 2008 Annual Conference* (Paper No. 08-197), S. 1–15.

Clément, Alain; Dorais, Martine; Vernon, Marcia (2008a): Multivariate approach to the measurement of tomato maturity and gustatory attributes and their rapid assessment by Vis-NIR spectroscopy. In: *Journal of agricultural and food chemistry* 56 (5), S. 1538–1544. DOI: 10.1021/jf072182n.

Clément, Alain; Dorais, Martine; Vernon, Marcia (2008b): Nondestructive measurement of fresh tomato lycopene content and other physicochemical characteristics using visible-NIR spectroscopy. In: *Journal of agricultural and food chemistry* 56 (21), S. 9813–9818. DOI: 10.1021/jf801299r.

Di Mascio, Paolo; Kaiser, Stephan; Sies, Helmut (1989): Lycopene as the Most Efficient Biological Carotenoid Singlet Oxygen Quencher. In: *Archives of biochemistry and biophysics* 274 (2), S. 532–538.

dos Santos, Cláudia A. Teixeira; Lopo, Miguel; Páscoa, Ricardo N. M. J.; Lopes, João A. (2013): A review on the applications of portable near-infrared spectrometers in the agro-food industry. In: *Applied spectroscopy* 67 (11), S. 1215–1233. DOI: 10.1366/13-07228.

D'Souza, Mervy C.; Singha, Suman; Ingle, Morris (1992): Lycopene Concentration of Tomato Fruit can be Estimated from Chromaticity Values. In: *HortScience* 27 (5), S. 465–466.

Felix Instruments (2017): F-750 Produce Quality Meter. Specifications. Online verfügbar unter https://felixinstruments.com/food-science-instruments/portable-nir-analyzers/f-750-produce-quality-meter/.

Fish, Wayne W.; Perkins-Veazie, Penelope; Collins, Julie K. (2002): A Quantitative Assay for Lycopene That Utilizes Reduced Volumes of Organic Solvents. In: *Journal of Food Composition and Analysis* 15 (3), S. 309–317. DOI: 10.1006/jfca.2002.1069.

Gautier, Hélène; Diakou-Verdin, Vicky; Bénard, Camille; Reich, Maryse; Buret, Michel; Bourgaud, Frédéric et al. (2008): How does tomato quality (sugar, acid, and nutritional quality) vary with ripening stage, temperature, and irradiance? In: *Journal of agricultural and food chemistry* 56 (4), S. 1241–1250. DOI: 10.1021/jf072196t.

Goisser, Simon; Krause, Julius; Fernandes, Michael; Mempel, Heike (2019a): Determination of tomato quality attributes using portable NIR-sensors. In: *OCM 2019 - Optical Characterization of Materials: Conference Proceedings*, S. 1–12. DOI: 10.5445/IR/1000092314.

Goisser, Simon; Mempel, Heike; Bitsch, Vera (2019b): Potential Applications of Food-Scanners in Fruit and Vegetable Supply Chains and Possible Consequences for the German Market. In: *Proceedings in System Dynamics and Innovation in Food Networks*, S. 173–181. DOI: 10.18461/pfsd.2019.1917.

Hyman, Joshua R.; Gaus, Jessica; Foolad, Majid R. (2004): A Rapid and Accurate Method for Estimating Tomato Lycopene Content by Measuring Chromaticity Values of Fruit Purée. In: *Journal of the American Society for Horticultural Science* 129 (5), S. 717–723.

Kaur, Harpreet; Künnemeyer, Rainer; McGlone, Andrew (2017): Comparison of hand-held near infrared spectrophotometers for fruit dry matter assessment. In: *Journal of Near Infrared Spectroscopy* 25 (4), S. 267–277. DOI: 10.1177/0967033517725530.

Kugler, Friedrich (2009): Rot und gesund. Bei Bötsch Gemüsebau in Salmsach werden besonders geschmacksintensive Tomaten mit hohem Lycopingehalt kultiviert. In: *Migros Magazin (Ostschweiz)* (25), S. 85. Online verfügbar unter www.boetsch-gemuese.ch.

Kusumiyati; Akinaga, Takayoshi; Tanaka, Munehiro; Kawasaki, Sheishi (2008): On-tree and after-harvesting evaluation of firmness, color and lycopene content of tomato fruit using portable NIR spectroscopy. In: *Journal of Food, Agriculture & Environment* 6 (2), S. 327–332.

Li, Mo; Qian, Zhiqing; Shi, Bingwen; Medlicott, Jane; East, Andrew (2018): Evaluating the performance of a consumer scale SCiO™ molecular sensor to predict quality of horticultural products. In: *Postharvest Biology and Technology* 145, S. 183–192. DOI: 10.1016/j.postharvbio.2018.07.009.

Li, Tianhua; Zhong, Chongzhe; Lou, Wei; Wei, Min; Hou, Jialin (2017): Optimization of Characteristic Wavelengths in Prediction of Lycopene in Tomatoes Using Near-Infrared Spectroscopy. In: *Journal of Food Process Engineering* 40 (1), e12266. DOI: 10.1111/jfpe.12266.

Rateni, Giovanni; Dario, Paolo; Cavallo, Filippo (2017): Smartphone-Based Food Diagnostic Technologies: A Review. In: *Sensors (Basel, Switzerland)* 17 (6). DOI: 10.3390/s17061453.

Rijk Zwaan (2019): Lyterno RZ F1 (72-471) - Product information. Hg. v. Rijk Zwaan. Online verfügbar unter https://www.rijkzwaan.de/ihre-sorte/tomate/lyterno-rz, zuletzt geprüft am 19.02.2020.

Sadler, G.; Davis, J.; Dezman, D. (1990): Rapid Extraction of Lycopene and β-Carotene from Reconstituted Tomato Paste and Pink Grapefruit Homogenates. In: *Journal of Food Science* 55 (5), S. 1460–1461.

Seren, Soley; Liebermann, Ronald; Bayraktar, Ulas D.; Heath, Elisabeth; Sahin, Kazim; Andic, Fundagul; Kucuk, Omer (2008): Lycopene in Cancer Prevention and Treatment. In: *American Journal of Therapeutics* 15, S. 66–81.

Sheng, Ren; Cheng, Wu; Li, Huanhuan; Ali, Shujat; Akomeah Agyekum, Akwasi; Chen, Quansheng (2019): Model development for soluble solids and lycopene contents of cherry tomato at different temperatures using near-infrared spectroscopy. In: *Postharvest Biology and Technology* 156, S. 110952. DOI: 10.1016/j.postharvbio.2019.110952.

Sies, Helmut; Stahl, Wilhelm (1995): Vitamin E and C, b-carotene and other carotenoids as antioxidants. In: *The American Journal of Clinical Nutrition* 62, 1315S-1321S.

Stahl, Wilhelm; Heinrich, Ulrike; Wiseman, Sheila; Eichler, Olaf; Sies, Helmut; Tronnier, Hagen (2001): Dietary Tomato Paste Protects against Ultraviolet Light-Induced Erythema in Humans. In: *Journal of Nutrition* 131 (5), S. 1449–1451. DOI: 10.1093/jn/131.5.1449.

Szuvandzsiev, Péter; Helyes, Lajos; Lugasi, Andrea; Szántó, Csongor; Baranowski, Piotr; Pék, Zoltán (2014): Estimation of antioxidant components of tomato using VIS-NIR reflectance data by handheld portable spectrometer. In: *International Agrophysics* 28 (4). DOI: 10.2478/intag-2014-0042.

Tilahun, Shimeles; Park, Do Su; Seo, Mu Hong; Hwang, In Geun; Kim, Seok Hyeon; Choi, Han Ryul; Jeong, Cheon Soon (2018): Prediction of lycopene and β-carotene in tomatoes by portable chroma-meter and VIS/NIR spectra. In: *Postharvest Biology and Technology* 136, S. 50–56. DOI: 10.1016/j.postharvbio.2017.10.007.

USDA (1975): USDA Color Chart. Color Classification Requirements in Tomatoes. Unter Mitarbeit von United Fresh Fruit and Vegetable Association, Agricultural Marketing Service und Fruit and Vegetable Division. The John Henry Company. Lansing, Mich. Online verfügbar unter https://ucanr.edu/repository/view.cfm?article=83755%20&groupid=9.

Wiedemair, Verena; Huck, Christian W. (2018): Evaluation of the performance of three hand-held near-infrared spectrometer through investigation of total antioxidant capacity in gluten-free grains. In: *Talanta* 189, S. 233–240. DOI: 10.1016/j.talanta.2018.06.056.

Tables

Table 1: Operating parameters of the three portable VIS/NIR devices used in this study

Instrument	Spectrometer	Light source	Wavelength range (nm)	Resolution (nm)	Sampling interval (nm)	Measurement time[b] (s)
F-750 Produce Quality Meter	Carl Zeiss MMS-1	Xenon tungsten lamp	310 - 1100	8 - 13	3.3	10
H-100F	Enhanced CMOS	Halogen lamp	650 - 950	< 20[a]	2	2
SCiO™ Molecular Sensor	Image sensor	LED	740 - 1070	< 1	1	8

[a]Estimated by Kaur et al. (2017)

[b]Based on own time recordings

Table 2: Average values of lycopene content and color readings across different ripening stages of ripening

Ripening stage [a,b]	Lycopene content (mg kg^{-1})	L^*	a^*	b^*	a^*/b^*	$(a^*/b^*)^2$	TCI
Green	0.84 (± 0.88)e	48.01 (± 2.52)a	-6.72 (± 0.96)f	26.11 (± 2.70)bc	-0.26 (± 0.02)g	0.07 (± 0.01)e	-10.43 (± 1.30)f
Breaker	4.94 (± 1.99)e	42.46 (± 1.96)b	-2.80 (± 1.08)e	26.81 (± 2.44)bc	-0.10 (± 0.04)f	0.01 (± 0.01)e	-4.86 (± 1.71)e
Turning	14.78 (± 2.79)e	39.99 (± 1.41)c	4.35 (± 2.62)d	28.63 (± 2.41)ab	0.15 (± 0.09)e	0.03 (± 0.03)e	7.56 (± 4.34)d
Pink	44.74 (± 12.66)d	36.48 (± 1.73)d	17.29 (± 3.47)c	30.99 (± 2.65)a	0.56 (± 0.12)d	0.33 (± 0.13)d	26.71 (± 5.37)c
Red / Harvest	118.63 (± 21.37)c	32.27 (± 1.46)e	26.43 (± 1.86)b	30.15 (± 3.69)a	0.89 (± 0.10)c	0.80 (± 0.19)c	41.07 (± 3.94)b
8 DAH	152.67 (± 11.89)a	31.65 (± 0.72)e	26.51 (± 1.87)ab	28.82 (± 3.53)ab	0.93 (± 0.10)bc	0.87 (± 0.19)bc	42.87 (± 2.75)ab
13 DAH	153.90 (± 15.69)a	31.97 (± 0.60)e	26.83 (± 2.17)ab	26.86 (± 2.87)bc	1.00 (± 0.06)ab	1.01 (± 0.11)ab	44.27 (± 1.36)a
17 DAH	145.35 (± 9.52)ab	32.50 (± 0.83)e	26.83 (± 2.17)ab	26.19 (± 2.81)bc	1.03 (± 0.06)a	1.06 (± 0.11)a	44.08 (± 1.16)a
20 DAH	139.83 (± 9.81)ab	32.43 (± 0.75)e	25.92 (± 2.66)b	24.59 (± 3.11)c	1.06 (± 0.05)a	1.12 (± 0.11)a	44.79 (± 1.10)a
22 DAH	133.74 (± 9.84)b	32.26 (± 0.64)e	28.97 (± 1.91)a	27.71 (± 1.71)ab	1.05 (± 0.04)a	1.09 (± 0.08)a	44.79 (± 1.04)a

[a]Values in parenthesis represent standard deviations

[b]Values in the same row indicated with different letters are significantly different at $P \leq 0.05$ using analysis of variance and Tukey's multiple comparison test

Table 3: Linear and exponential regressions and Pearson correlations between color readings and lycopene content of tomatoes

Color value	Linear regression R^2	Exponential regression R^2	Pearson's correlation coefficient	Equation yielding the best correlation for lycopene (mg kg^{-1})
*L**	0.79	0.94	-0.89	$4000000\ e^{-0,32\ L^*}$
*a**	0.85	0.90	0.92	$4,002\ e^{0,1323\ a^*}$
*b**	0.00	0.01	-0.03	$8,2324\ e^{0,0608\ b^*}$
a/b**	0.88	0.90	0.94	$4,2785\ e^{3,5775\ (a^*/b^*)}$
$(a^*/b^*)^2$	0.87	0.68	0.93	$128,41\ (a^*/b^*)^2 + 9,5216$
TCI	0.88	0.91	0.94	$3,9269\ e^{0,0828\ TCI}$

Table 4: Performance of three commercially available food-scanners in predicting lycopene content (mg/kg) of intact tomato fruit

Device	n	r^2_C	$RMSE_C$	r^2_{CV}	$RMSE_{CV}$	PC
F-750 Produce Quality Meter (λ = 477 - 1059 nm)	165	0.96	12.06	0.96	12.29	2
H-100F (λ = 650 - 950 nm)	165	0.95	13.87	0.95	14.28	4
SCiO™ Molecular Sensor (λ = 740 - 1070 nm)	165	0.94	15.28	0.92	17.07	9

r^2_C: coefficient of determination in calibration; $RMSE_C$: root mean square error of calibration; r^2_{CV}: coefficient of determination in cross validation; $RMSE_{CV}$: root mean square error of cross validation; PC: principal components

8 Food-scanner applications in the fruit and vegetable sector

In: Landtechnik 76 (1), p. 52-67

Cite as: Goisser, S.; Wittmann, S.; Mempel, H. (2021): Food-scanner applications in the fruit and vegetable sector. Landtechnik 76 (1), p. 52-67.

DOI: https://doi.org/10.15150/lt.2021.3264

LANDTECHNIK 76(1), 2021, 52-67
DOI:10.15150/lt.2021.3264

Food-scanner applications in the fruit and vegetable sector

Simon Goisser, Sabine Wittmann, Heike Mempel

In the past few years, portable and smartphone-based diagnostic technologies have found their way into the agri-food industry. The aim of this research was to evaluate the performance of portable near-infrared (NIR) spectrometers, so called food-scanners, with regard to their predictive accuracy of important quality parameters of fruit and vegetables. Food-scanner measurements were performed in combination with destructive measurements of the corresponding quality trait (sugar content, dry matter, relative water content) on a wide range of produce from the fruit and vegetable assortment. This study evaluated dry matter content of apple, avocado, blueberry, table grape and tangerine, which yielded cross validation results (r^2) of up to 0.95, 0.87, 0.94, 0.92 and 0.92 respectively. Furthermore, the evaluation of food-scanner spectra for the prediction of sugar content of blueberry, kiwi, mango, persimmon, table grape, tangerine and tomato yielded cross validations (r^2) of up to 0.95, 0.84, 0.80, 0.75, 0.95, 0.93, and 0.87. Furthermore, relative water content of ginger obtained a cross validation correlation of r^2 = 0.91. The results show that these traits can be predicted with a high degree of accuracy using non-destructive measurements performed with three commercially available food-scanners SCiO^{TM}, F-750 Produce Quality Meter, and H-100F. Consequently, food-scanners can be used as objective measurement tools along the supply chain of fresh produce to quickly determine fruit quality. In addition, a practical example shows the potential of these instruments for non-destructive quality assessment in incoming goods control at fruit and vegetable wholesalers over a time period of several weeks. Furthermore, possible areas of application of food-scanners along the supply chain of fresh produce are discussed, possibilities for practical applications are presented and time-saving means are highlighted.

Keywords
NIR spectroscopy, food-scanner, non-destructive measurement, quality control

Near-infrared (NIR) technology has many specialist areas of application today. A common application is in the control of chemical, petrochemical and pharmaceutical products. Within the chemical industry, NIR spectrometers are also used in process control and for incoming goods inspection of raw materials. Another area of NIR application is waste separation, where a wide variety of composites and types of plastic are detected and separated through sorting lines. Furthermore, NIR technology is used in process control in the food industry and in the quality analysis of agricultural products (Pasquini 2003). Within these fields of application and in addition to the control of products for human consumption, quality determination of raw materials is of particular importance. NIR spectrometers therefore are used for quality measurements of cereals, flour, and milk. Further raw products such as feed are analyzed for moisture, protein and raw fiber content using these devices. The technology is also suitable for determining the water content in numerous agricultural products (Cozzolino 2009). Various pre-

received 15 August 2020 | accepted 23 December 2020| published 24 March 2021 vious experiments on horticultural produce illustrated the potential of using NIR spectroscopy for the prediction of important fruit quality parameters, e.g. sugar content and acidity in apples, oranges, kiwifruit, peaches, as well as kiwifruit dry matter and firmness of peaches and kiwifruit (WANG et al. 2015). Within the last few years, mobile smartphone-based diagnostic technologies have found their way into the agri-food sector. Those devices could replace expensive and bulky laboratory equipment in the near future. Various start-up companies are involved in the development of these measurement tools (RATENI et al. 2017). In particular, the technology of miniaturized and portable NIR spectroscopy has great potential in various areas of application due to its non-destructive use and the link to machine-learning algorithms. Individual companies are already targeting applications in the food industry. They refer to their spectrometers as food-scanners, which provide information on the nutrient content, allergens and contaminations and are intended to determine relevant information on food fraud, food counterfeiting and food quality within a few seconds (SPECTRAL ENGINES 2020, Tellspec 2020). Other companies are focusing slightly larger yet portable NIR devices specifically designed for the measurement of internal quality parameters such as sugar content and dry matter of various kind of fruit (FELIX INSTRUMENTS 2020, SUNFOREST 2020). First scientific studies, which employed these portable and miniaturized instruments, indicate good performance and prediction results for a variety of produce and quality parameters, e.g., dry matter content of apple, kiwifruit and stone fruit (KAUR et al. 2017), dry matter and oil content of avocado (NCAMA et al. 2018), dry matter and sugar content of mango (SANTOS NETO et al. 2017), and sugar content of pear (CHOI et al. 2017).

A recent qualitative study conducted with actors of the German fresh produce supply chain highlights the potential that these food-scanners could unlock by facilitating future quality control on various levels along the supply chain through non-destructive, objective and fast measurements (GOISSER et al. 2020a).

Measurement procedure and data collection of important quality attributes within the assortment of fruit and vegetables

In order to evaluate the applicability of these new types of devices within the practice of fruit and vegetable trade, food-scanners were tested on a large range of available produce within the fruit and vegetable retail sector. Additional fruit and vegetables, of which the application of NIR spectrometers for the prediction of quality traits have not yet been thoroughly researched in the scientific literature (e.g., blueberry, ginger), were analyzed during the course of these studies. Further special features, such as the prediction of relevant quality parameters of thick-skinned fruit (e.g., mango, avocado, citrus fruit) as well as the non-destructive measurement through packaging film were taken into account. Experiments focused on the two quality parameters sugar content and dry matter, as these quality traits often serve as indicators for ripeness and eating quality of fruit. In case of ginger, relative water content of tubers was examined as indicator of freshness.

Fruit analyzed during the course of this research are listed in Table 1. Batches of different size of apple, avocado, blueberry, kiwi, mango, persimmon, table grape, and tangerine were purchased from local retail stores as well as fruit wholesale and analyzed at laboratories of the University of Applied Sciences Weihenstephan-Triesdorf in Freising, Germany. In order to obtain the widest possible range of the respective quality parameters of interest, care was taken to integrate different fruit qualities when produce was procured (Table 1). Additional measurements for the determination of quality parameters through packaging film of table grapes and tomatoes were performed during incoming goods control of a German fruit and vegetable trading company. For the measurement of ginger, fresh plant material of a local plant nursery was obtained and transported to University facilities for subsequent analysis.

During the course of these studies, three commercially available miniaturized and hand-held NIR spectrometers were used: the SCiO™ (version 1.2, Consumer Physics, Hod HaSharon, Israel); the F-750 Produce Quality Meter (Firmware v.1.2.0 build 7041, Felix Instruments, Portland, USA) and the H-100F (Sunforest, Incheon, Korea). All three devices measure in direct contact with fruit samples and utilize interactance mode, therefore only light transmitted through the sample tissue is detected and more information about the actual sample composition and constituents is obtained (Pasquini 2003). The obtained spectra are used to calculate absorbance spectra, which then is utilized for further analysis. The detailed characteristics and operating parameters of these food-scanners, such as spectral resolution and sampling interval, have been elaborately described by Kaur et al. (2017).

Table 1: Products and respective quality parameters examined in this study including reference methods and the source of quality variation

Product	Quality parameter	Reference method	Source of variation in quality
Apple	Dry matter	Drying oven	Integration of various cultivars
Avocado	Dry matter	Drying oven	Combination of different fruit calibres and origins
Blueberry	Dry matter / Brix	Drying oven / refractometer	Pooling of various origins and growing methods (traditional / organic)
Ginger	Water content	Weight loss measurement	Examination of freshly harvested ginger during convective drying
Kiwi (green & gold)	Brix	Refractometer	Merging of fruit with differences in external appearance
Mango	Brix	Refractometer	Combination of air freighted and traditionally transported fruit
Persimmon	Brix	Refractometer	Pooling of various origins and growing methods (traditional / organic)
Table grape (green)	Dry matter / Brix / Brix through PE film	Drying oven / refractometer	Incorporation of various cultivars and origins
Tangerine	Dry matter / Brix	Drying oven / refractometer	Integration of various cultivars
Tomato	Brix through PE film	Refractometer	Combination of various cultivars, types (cherry, salad) and origins

In order to develop NIR prediction models, the first step was the collection of NIR fruit spectra using the respective food-scanner. In contrast to measurements under laboratory conditions, all experiments were performed under condition as realistic and practically relevant as possible. Various factors, such as differences in ambient and fruit temperature (e.g., 4 °C - 22 °C) and variations in lighting conditions, which are often found in incoming goods control of wholesalers and retailers, were therefore integrated into these prediction models. For mango, avocado, tangerine, tomato, apple, kiwi and persimmon, the exact location of spectra collection was used for subsequent reference measurement, whereas for table grapes, blueberries and ginger one spectra per specimen was taken and the whole berry/tuber was used for reference measurement afterwards. For measurements through packaging film, small pre-cut polyethylene (PE) film samples were placed between fruit sample and food-scanner during the recording of spectra.

Spectra collected with the F-750 was transferred from the SD-Card to a computer for further chemometrical analysis using the associated Model Builder Software (version 1.3.0.192 BETA). According to the manufacturer, the software applies non-linear iterative partial least squares (NIPALS) regression and uses the leave-one-out method for cross validation. Absorbance spectra in second derivative form in the wavelength range 729 - 975 nm was utilized for building of prediction models, since these settings are used as standard by the software.

Similarly, absorbance spectra recorded with the SCiO™ was correlated to fruit quality parameters using the cloud-based browser application The Lab (Consumer Physics). The cloud application's default pre-processing settings of spectra were used in this study, namely logarithm, averaging all four scans per sample, first derivative (window of 35 and polynom degree of 2), selecting the full wavelength range from 740 - 1070 nm and subtracting the average. The application uses every tenth spectra for cross-validation. The algorithm was set to partial least square regression (PLSR) and additional options for outlier detection as well as additional filters were disabled.

For the H-100F, recorded spectra were exported from the Sunforest H-100 Laboratory Programm (Sunforest). The device provides absorbance spectra in second derivative form in the wavelength range from 650 - 950 nm. Due to the lack of evaluation software provided by the manufacturer, spectra were analyzed with the multivariate data analysis software The Unscrambler® (version 10.5.1, Camo, Oslo, Norway). The H-100F was only used for the evaluation of tangerines, and the two scans per fruit were averaged prior to multivariate analysis. Following the procedure of the F-750, no further pre-processing was applied, and 20 random samples were used for cross-validation.

Coefficient of correlation after cross-validation (r^2_{CV}) as well as the root mean square error after cross-validation ($RMSE_{CV}$) derived from the respective software were used for performance evaluation of the various prediction models (Table 2).

The determination of reference values of the respective quality parameters was performed immediately after collection of fruit spectra. Sugar content in terms of Brix was analyzed by cutting out the area of spectra collection (in case of table grapes and blueberries the whole berry was used), squeezing the fruit tissue with a garlic press and measuring the mixed juice with a digital refractometer (HI 96801, Hanna Instruments, Woonsocket, USA). Dry matter of the area of spectra collection was measured gravimetrically (for table grapes and blueberries whole berries were analyzed). Fresh fruit tissue was weighed and subsequently dried in an oven at 45 °C (avocado) and 80 °C (table grape, blueberry, tangerine, apple) for 48 h. The final dry weight was used for calculation of dry matter as the percentage of dry weight to initial wet weight of each specimen. In order to determine the relative water content of ginger, small tubers fresh from the nursery were examined. Measurement was performed on five consecutive days during which tubers were stored at room temperature (19.8 °C ± 0.9 °C) and relative humidity of 48% ± 3%. After collection of food-scanner spectra, tubers were weighed, and the current weight noted. At the end of day five, dry weight of tubers was determined gravimetrically by drying the tubers in an oven at 80 °C for 48 h. Relative water content was then calculated using the equation provided by CECCATO et al. (2001), on the assumption that fresh tubers showed full turgor weight on day one:

$$Relative\ water\ content = \frac{Tuber\ weight\ on\ the\ respective\ day - oven\ dry\ weight}{Tuber\ turgor\ weight\ on\ day\ one - oven\ dry\ weight}$$

Performance of prediction models for fruit quality parameters

Prediction results with respect to r^2_{CV} and $RMSE_{CV}$ obtained from multivariate analysis for every quality parameter and respective food-scanner are listed in Table 2.

Overall, the correlation of spectra with sugar content of fruit yielded good results. Prediction models of high accuracy ($r^2_{CV} > 0.90$) were obtained for blueberries (Figure 1A) and table grapes. Likewise, the measurement through thick-skinned fruit such as tangerines provided very good results. Further good correlations of sugar content were obtained for mango and kiwi (r^2_{CV} = 0.78 - 0.84), whereas the prediction of persimmon sugar content yielded better results using the F-750 (r^2_{CV} = 0.75) compared to the SCiO™ (r^2_{CV} = 0.63). Nevertheless, all values of root mean square errors ($RMSE_{CV}$), which serve as indicator of prediction error, ranged from 0.36 to 0.92 °Brix and can therefore be deemed acceptable for a non-destructive evaluation of fruit quality. Results from this study are comparable to findings reported in the literature, such as the determination of sugar content of kiwi (McGlone and Kawano 1998), mango (Schmilovitch et al. 2000), table grapes (Donis-González et al. 2020) and tangerines (McGlone et al. 2003a) using NIR spectrometers.

Table 2: Performance of selected food-scanners in predicting internal fruit quality of various types of fruit using PLS algorithm. r^2_{CV}: r2 of cross-validation; $RMSE_{CV}$: root mean square error of cross-validation; PC: principal components

Product	Parameter	F-750	SCiO™	H-100F	n	r^2_{CV}	$RMSE_{CV}$	PC
Apple[1]	Dry matter	X			100	0.95	0.53	6
Avocado[2]	Dry matter	X			195	0.85	1.50	6
	Dry matter		X		195	0.87	1.42	16
Blueberry[1]	Brix	X			275	0.90	0.68	9
	Brix		X		275	0.95	0.47	12
	Dry matter	X			100	0.92	0.59	7
	Dry matter		X		100	0.94	0.51	6
Ginger[2]	Relative water content	X			150	0.91	6.44	5
Kiwi (green)[1]	Brix	X			120	0.84	0.63	8
Kiwi (gold)[1]	Brix	X			120	0.78	0.80	7
Mango[1]	Brix	X			140	0.80	0.58	6
Persimmon[1]	Brix	X			45	0.75	0.76	5
	Brix		X		45	0.63	0.92	3
Table grape (green)[1]	Brix		X		80	0.94	0.51	7
	Brix	X			150	0.95	0.57	5
	Brix (through PE film)	X			75	0.93	0.65	7
	Dry matter		X		80	0.92	0.55	7
	Dry matter	X			160	0.92	0.46	4
Tangerine[1]	Brix			X	80	0.93	0.36	6
	Dry matter			X	174	0.92	0.53	5
Tomato[1]	Brix (through PE film)	X			177	0.87	0.57	7

1) Various cultivars used for model building 2) Individual cultivars used for model building

Prediction models for sugar content in persimmon obtained the lowest accuracy in our study. These results are inferior to previous literature reports (JANNOK et al. 2014). Additionally, we examined the predictive capability of table grape sugar content through packaging film with the F-750 and obtained prediction models similar to those without packaging (Figure 1B). Similarly, the determination of tomato sugar content through packaging film yielded prediction models of comparable performance to those reported in earlier experiments (GOISSER et al. 2018). A thorough literature review showed that the prediction of internal quality parameters of fruit through packaging material has not yet been studied.

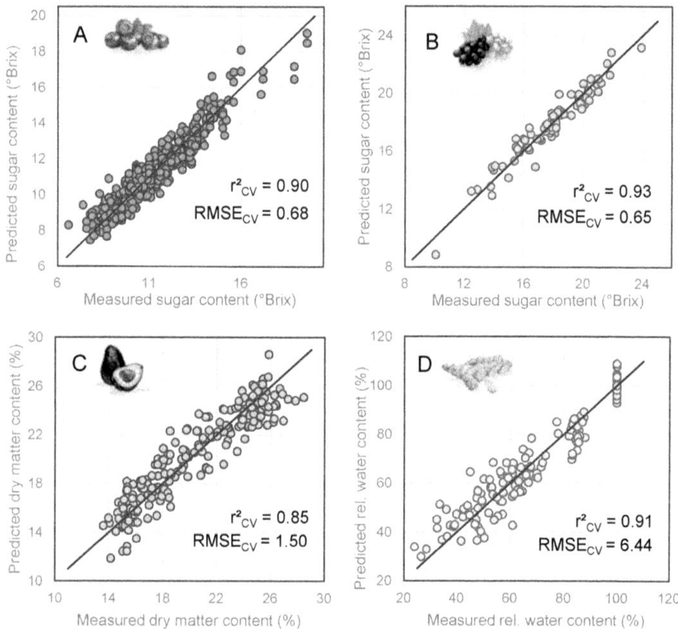

Figure 1: Correlation between NIR food-scanner spectra and quality parameter for A) sugar content of blueberries; B) sugar content of grapes through packaging film; C) dry matter content of avocado; D) relative water content of ginger

Prediction models developed for dry matter content in various types of fruit yielded good results (r^2_{CV} = 0.85 - 0.95). Models of high accuracy could be derived for dry matter of apples, blueberries, table grapes and tangerines (r^2_{CV} > 0.92; $RMSE_{CV}$ = 0.46 - 0.59). Our prediction models derived for dry matter in apples are superior to findings of MØLLER et al. (2013) and similar to results of MCGLONE et al. (2003b). Results from our experiments yielded better correlations for the prediction of table grape dry matter compared to findings reported in the literature (DONIS-GONZÁLEZ et al. 2020), whereas the performance for dry matter prediction in intact tangerines was analogous to results of GUTHRIE et al. (2005). The portable devices used in our study performed good in predicting dry matter of blueberries (r^2_{CV} > 0.92). However, no comparable reports were found for dry matter prediction of blueberries within the scientific literature. The prediction of dry matter content of avocado (Figure 1C) resulted in models of slightly lower accuracy (r^2_{CV} = 0.85 - 0.87) and higher prediction errors ($RMSE_{CV}$ = 1.42 - 1.50). This slight deviation in model accuracy compared other fruit could be explained by the dark skin of avocados, since a large part of NIR radiation is intercepted, which results in loss of spectral information. Nevertheless, our findings are in line with results achieved by OLAREWAJU et al. (2016) for the prediction of avocado dry matter.

The determination of the relative water content of ginger (Figure 1D) obtained good results (r^2_{CV} = 0.91; $RMSE_{CV}$ = 6.44). Comparable findings have been reported by LI et al. (2011) by applying PLS models and using a laboratory spectrometer for the determination of moisture content in ginger.

For almost all quality parameters examined, in particular sugar content and dry matter of various important fruit types, the developed prediction models yielded a high and therefore practical accuracy. In most cases, relatively small sample sizes were used to create prediction models. With regard to the utilization of these devices along the fruit and vegetable supply chain, the creation of new and separate prediction models is easy and can be realized quickly. It should be noted that the number of principal components has been automatically chosen by the respective software, which in some cases (e.g. avocado dry matter and blueberry sugar content for SCiO^{TM}) resulted in a very high number of PLS factors. Additionally, the models were validated using cross-validation from the same sample set. Due to the variability within some of the presented prediction models as well as the high number of PCs, validation with external samples must be performed to guarantee model robustness for practical applications. This is especially important if quality is predicted for fruit varieties or fruit origins not yet present within the sample pool.

Validation in practice – utilizing a food-scanner for incoming goods control

In order to demonstrate the potential of food-scanners for the quality assessment during incoming goods control, a practical experiment was carried out. For this purpose, the F-750 Produce Quality Meter from Felix Instruments was selected in coordination with the fruit trader and used in two consecutive years within an incoming goods quality control department of a German fruit trader for the prediction of table grape sugar content. Table grapes are usually marketed as a type of produce with no particular focus on cultivars. Therefore, the GLOBALG.A.P. number (GGN) was used as principal indicator of grapes emanating from different producers and therefore of different orchard origin. The procedure is described in Figure 2, the performance of the various prediction models built during the experiment is shown in Table 3. In order to illustrate the time saved by using a food-scanner, the time required for destructive and NIR measurements was recorded and compared (Table 4).

In a first step, data was collected during three consecutive weeks of quality assessment in the spring of 2019. Within the first week, NIR spectra of ten green table grapes per day were recorded and the respective sugar content was measured destructively using a digital refractometer (HI 96801, Hanna Instruments, Woonsocket, USA). Using the resulting 50 reference values and fruit spectra, a first prediction model (PM-1) was developed with the software provided by Felix Instruments (Model Builder Software version 1.3.0.192 BETA). During the second week of quality assessment, NIR predictions for sugar content of ten green table grapes were performed and the actual sugar content of each table grape was destructively determined with the refractometer. As illustrated by the comparative Boxplot in Figure 2 only slight deviations between F-750 predictions and destructive measurements of the overall daily quality could be detected.

Table 3: Performance of various prediction models built with the F-750 Produce Quality Meter and Model Builder Software for the prediction of table grape sugar content during the practical experiments. r^2_{CV}: r^2 of cross-validation; $RMSE_{CV}$: root mean square error of cross-validation; r^2_V: r^2 of validation by external samples; $RMSE_V$: root mean square error of validation by external samples

Model label	No. of cultivar	No. of GLOBALG.A.P. numbers	n	r^2_{CV}	$RMSE_{CV}$	r^2_V	$RMSE_V$
PM-1	2	3	50	0.87	0.74	0.75	0.84
PM-2	2	5	100	0.96	0.56	0.93	0.65
PM-3	2	9	150	0.95	0.57	-	-
PM-4	2	13	195	0.95	0.52	0.92	0.77
PM-5	2	19	345	0.93	0.62	-	-

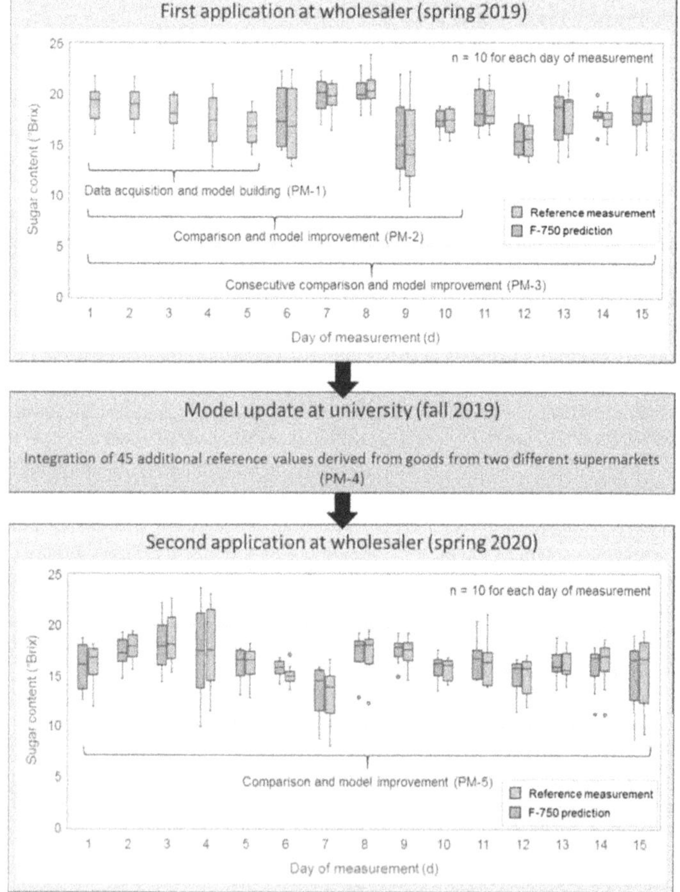

Figure 2: Procedure for utilizing the F-750 Produce Quality Meter for the non-destructive measurement of sugar content of green table grapes during incoming goods control at a German fruit wholesaler

The additional data collected during the second week was used for the improvement of the first prediction model (PM-1), which led to the building of a second prediction model (PM-2) using the provided software.

During the third week, the procedure from the second week was repeated. The comparison between F-750 prediction and destructive measurement showed high conformity (Figure 2). At the end of the third week of the first experiment part, all data was pooled in a third prediction model (PM-3).

In preparation for the next practical application, the third prediction model (PM-3) was updated at the University of Applied Sciences Weihenstephan-Triesdorf in the fall of 2019 by adding 45 additional spectra and reference values of green table grapes bought from two different local supermarkets. Data was again pooled and a fourth prediction model was developed (PM-4).

In the spring of 2020, the updated prediction model (PM-4) was again used during quality assessment of incoming goods at the same wholesale company for a second three-week practical experiment. Every day, fruit spectra, NIR predictions for sugar content and destructive reference measurements of ten green table grapes were performed and recorded. Analog to the first practical application, the NIR prediction model in 2020 was updated weekly by integrating the recorded spectra and reference values. As indicated by the second Boxplot in Figure 2, the second experiment initially showed only a slight deviation of prediction and reference values. The overall coverage of the daily fruit quality using non-destructive NIR predictions can be described as very satisfying. Furthermore, the continuous improvements resulted in an increase of conformity of NIR prediction values and real reference values towards the end of the three-week period. At the end of the second practical experiment, all data was aggregated in a final fifth model (PM-5). The models PM-1 and PM-2 developed in 2019 have been tested for model robustness using the data of the consecutive week for model validation. Similarly, PM-4 was validated using the data collected in 2020 (Table 3).

In an additional experiment the time needed to measure the sugar content of 25 table grapes via refractometer and food-scanner was recorded and compared (Table 4). An essential advantage of the NIR measurements is the fact that the measurements can be carried out non-destructively and directly on the working platform and therefore no additional working time is required for cleaning of equipment and transport to an adjacent laboratory. Furthermore, no additional equipment (e.g., paper towels, knife, cutting board, deionized water) is needed when measurements are performed via food-scanner compared to a refractometer.

Table 4: Comparison of the time required for sugar measurements using refractometer and NIR-device

Measurement procedure	Refractometer	F-750 Produce Quality Meter
Number of samples	25	25
Measurement time / sample (s)	30	15
Time for cleaning and transport / sample (s)	15	0
Total time (min)	18:45	6:15

The presented experiment illustrates the great potential of portable and miniaturized NIR food-scanners for a non-destructive quality assessment of internal fruit quality during incoming goods control. As shown on the example of table grapes, the creation of new prediction models is easy and can be realized quickly by utilizing the software provided by device manufacturers. Furthermore, the experiment demonstrated that prediction models of high accuracy built for specific type of fruit can be used at a later time while still showing high conformity with destructive measurements. Moreover, fruit of origin previously not included within the data pool, as illustrated through the growing number of new producers, could be predicted with high accuracy. Treating this new data as samples for external validation resulted in high model robustness for PM-2 and PM-4. In addition, the experiment shows that prediction models can be updated quickly and easily at any time by entering small new data sets. Compared to conventional destructive measurements, which are usually necessary to determine the respective quality parameters, food-scanners are comparatively easy to handle and demonstrated to save valuable working time. In addition, fewer produce has to be destroyed for quality assessment. The ability to measure non-destructively through the packaging film with high prediction accuracy, as indicated in previous experiments (Table 2) makes these devices attractive for use along the fruit and vegetable supply chain. Although only a practical validation for the prediction of table grape sugar content is shown in this study, we assume that it is similarly feasible for other fruits at reasonable expense. Thus, a practical application could become quite realistic for the future. In order to guarantee model robustness and prediction accuracy, a periodic validation and re-calibration is necessary.

Potential applications along the supply chain of fruit and vegetables

Based on our findings there are numerous possible applications for food-scanners along the fresh produce supply chain.

At the producer level, food-scanner can be used to monitor fruit quality during growth and maturing of fruit. Due to the portability of these devices, some of the relevant ripening parameters can be determined directly on the plant and harvest predictions can be derived at an early stage. Furthermore, harvesting operations in large orchards can be better coordinated. Certain manufacturers of portable NIR devices have specialized in this area of application and already offer devices specifically designed for various types of fruit (Felix Instruments 2020). Therefore, important ripening parameters such as sugar content and dry matter of tropical fruit (e.g., avocado, mango, kiwi) can be predicted in a non-destructive way. Once these products have been transported, food-scanners can be used to verify fruit quality at the next level of the supply chain. Some produce, especially fruit from overseas, are ripened in special ripening facilities just after arrival in the importing country. The possibility of non-destructive quality assessment enables fruit wholesalers to better control and organize these post-ripening processes.

During incoming goods control of fruit and vegetables at wholesale companies, defined quality criteria must be satisfied. For various types of fruit, statutory minimum requirements are specified (Table 5). Compliance with these standards and additional requirements imposed by retail chains is necessary for an efficient marketing of produce within the European market (UNECE 2019). However, these parameters can often not easily be determined, especially in the case of packaged produce. As stated in the latest draft of the OECD guidelines, NIR spectrometers comprise the ability to simultaneously predict various fruit quality parameters from a single NIR spectrum, therefore serving as "multidimensional predictors of consumer acceptance" (OECD 2018). The prediction results of various quality parameters using food-scanners in our studies indicate the potential of food-scanners, however further validations have to be performed. Following this, several quality parameters of fruit types provided with limited values could be predicted in a non-destructive way with high accuracy (e.g., avocado, kiwi, table grape, citrus). As we demonstrated for table grapes and tomatoes, there is great potential for the assessment of fruit quality through packaging film, which could facilitate incoming goods control even further.

Table 5: Overview of fresh fruit and vegetables with specific limits of fruit quality parameters according to the United Nations Economic Commission for Europe (UNECE 2019).

Produce	Limiting parameter	Respective document number (UNECE 2019)
Apple	Brix	FFV-50
Avocado	Dry matter	FFV-42
Citrus fruit	Brix, sugar/acid ratio, juice content	FFV-14
Kiwi	Brix, dry matter	FFV-46
Melon	Brix	FFV-23
Peach, Nectarine	Brix, firmness	FFV-26
Pineapple	Brix	FFV-49
Table grape	Brix, sugar/acid ratio	FFV-19
Watermelon	Brix	FFV-37

Portable NIR spectrometers can be integrated into the practical quality control process and the quality of produce can be examined on site directly on the pallet. On the one hand, workflow can be facilitated, which on the other hand saves valuable working time. Additionally, the application of food-scanners allows better objective, reliable and thus comparable quality recording in contrast to oftentimes subjective evaluations by test personnel (GOISSER et al. 2020a).

An extensive evaluation of the data that was recorded during incoming goods control of a German fruit and vegetable wholesaler shows the additional benefit of implementing NIR food-scanners. In this way, the daily recorded an already digitized values can be used to identify differences in delivered produce quality non-destructively, as illustrated by predictions of the F-750 Produce Quality Meter (Figure 3A). In order to protect the anonymity of the cooperation partner, the varieties are not listed in detail in the example shown. The comparison of the destructively determined and non-destructively measured sugar content values of table grapes during our incoming goods control experiment showed that the vast majority of the measured values (> 90%) were within an error range of ± 1 °Brix (Figure 3B). During incoming goods control at wholesalers, a reliable statement about the averaged sample is more relevant than the quality of a single product. Therefore, the accuracy determined in our experiments can be deemed appropriate for the determination of averaged values of a larger sample. Due to this measurement accuracy, fluctuations in the course of the season, individual supplier performance and quality differences depending on the region of origin can be recorded quickly. If necessary, appropriate measures (e.g., special advertising, discount campaigns) can be initiated.

Food-scanners can also be used for end-consumers as well as retail stores to distinguish entry-level products from premium products within the assortment of fruit and vegetables. Differences between these products are often primarily communicated through the layout and design of packaging. We used the data collected during the course of our incoming goods control project and combined the destructively recorded sugar contents of different commercial tomato varieties with the respective retail prices (Figure 3C). As illustrated, fruit qualities that can be clearly distinguished from one another are already offered at different prices. Oftentimes, these differences in internal fruit quality are not visible for unexperienced end-consumers, which makes the different pricing strategies difficult to understand. By using food-scanners in retail stores, these differences in internal fruit quality can be communicated to the end-consumer, e.g. by demonstrating live measurements in combination with sensory tasting of fruit. Dynamic pricing strategies, as described by DUAN and LIU (2019), which are based on a transparent fruit quality, could help retailers to profit from high-quality produce in the future. However, handling of these devices by unexperienced consumers is a challenge and must be taken carefully into account. Incorrectly performed measurements and false results could cause severe brand damage for retail markets and producers and lead to unnecessary food waste and recalls (PÖPPING and BOURDICHON 2018).

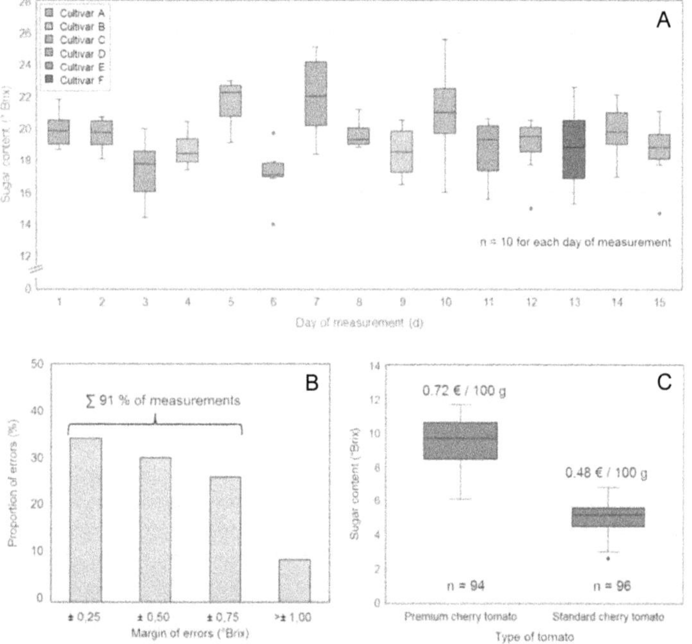

Figure 3: (A) Differences in sugar content of table grapes measured on 15 days during incoming goods control using the F-750 Produce Quality Meter; (B) margin of errors for food-scanner measurements compared to destructive measurements; (C) distinction of different types of tomato due to their sugar content and retail price (according to data from 2020, calendar week 33) using the F-750 Produce Quality Meter

Figure 4 summarizes the possible applications of food-scanners along the fresh produce supply chain. In order to allow a reasonable handling by end-consumers, additional aspects such as handling by inexperienced people and the correct interpretation of the displayed measurements must be taken into account (PÖPPING and BOURDICHON 2018). With regard to future application of these devices, the estimation of shelf life of fruit and vegetable became the focus of attention within the last few years. First studies conducted on selected produce such as apple (CORREA et al. 2015), asparagus (SÁNCHEZ et al. 2009) or lettuce (Giovenzana et al. 2014) indicate the possibility of modelling the shelf life of produce by using NIR spectrometry. Another future area of food-scanner application could be the prediction of sensory taste patterns. The results of initial studies on the combination of NIR spectra with sensory analysis on apples and grapes indicate that taste patterns and consumer preferences can be recognized using NIR spectroscopy (Mehinagic et al. 2003, Parpinello et al. 2013).

With regard to the development of new prediction models for the utilization of these devices in daily practice, as illustrated by the presented case study, a broad range of fruit quality and respective reference values must be considered in order to retain good prediction models. As demonstrated in previous studies (Wedding et al. 2013, Fan et al. 2019), the integration of biological and seasonal variability of fruit can help to build more robust and reliable prediction models and thereby unlock a great potential of non-destructive quality assessment via NIR food-scanners in practical applications for fresh produce.

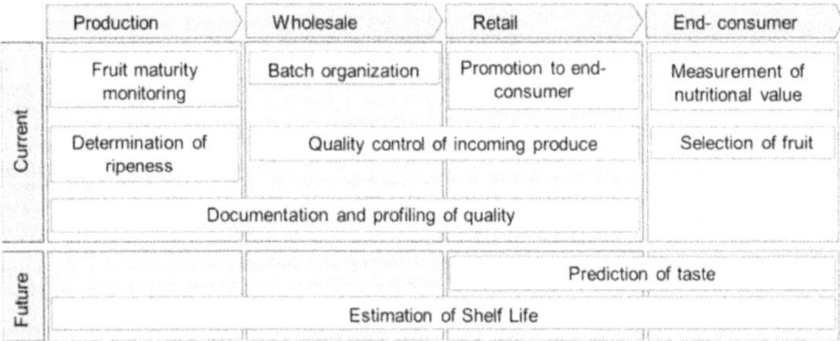

Figure 4: Areas of food-scanner applications along the fresh produce supply chain in current and future quality control

Conclusions

Our study illustrated the possible application of food-scanner at various levels of the fresh produce supply chain. Measurements made possible by food-scanners add value to the quality assessment within the trade of fruit and vegetables. Therefore, the quality of fresh produce can be communicated and tracked quickly and objectively along the supply chain. Our study also shows that further simplifications with regard to the practical use of these devices will be possible in the future. As demonstrated, food-scanners can be used for the building of various prediction models from data collected during incoming goods control or measuring fruit quality directly through packaging film. With regard to the use and benefit of these devices for end-consumers, further applications, such as the prediction of taste, are of great interest. Food-scanners could help to cluster the sensory quality of fruit into taste profiles (e.g., sweet, sour, fruity), which overall enables both retailers and end-consumers to select the desired quality in a targeted manner.

With regard to future investigations, the combination of information derived from NIR spectroscopy with additional data such as temperature and humidity during storage could constitute a new method of shelf life estimation of fresh produce. Using an improved assessment of shelf life at an early stage, the flow of produce along the supply chain can be controlled more specifically. Consequently, produce not suitable for the fresh market can be sent to alternative paths of food processing (e.g., smoothies, juices, dried produce), which helps saving valuable resources and avoids redundant food loss.

References

Ceccato, P.; Flasse, S.; Tarantola, S.; Jacquemoud, S.; Grégoire, J.-M. (2001): Detecting vegetation leaf water content using reflectance in the optical domain. Remote Sensing of Environment 77, 22-33. https://doi.org/10.1016/S0034-4257(01)00191-2

Choi, J.-H.; Chen, P.-A.; Lee, B.; Yim, S.-H.; Kim, M.-S.; Bae, Y.-S.; Lim, D.-C.; Seo, H.-J. (2017): Portable, non-destructive tester integrating VIS/NIR reflectance spectroscopy for the detection of sugar content in Asian pears. Scientia Horticulturae 220, 147-153. https://doi.org/10.1016/j.scienta.2017.03.050

Correa, A.R.; Quicazán, M.C.; Hernandez, C.E. (2015): Modelling the Shelf-life of Apple Products Accoring to their Water Activity. Chemical Engineering Transactions 43, 199-204. https://doi.org/10.3303/CET1543034

Cozzolino, D. (2009): Near infrared spectroscopy in natural products analysis. Planta medica 75, 746-756. https://doi.org/10.1055/s-0028-1112220

Donis-González, I.R.; Valero, C.; Momin, M.A.; Kaur, A.C.; Slaughter, D. (2020): Performance Evaluation of Two Commercially Available Portable Spectrometers to Non-Invasively Determine Table Grape and Peach Quality Attributes. Agronomy 10, 148. https://doi.org/10.3390/agronomy10010148

Duan, Y.; Liu, J. (2019): Optimal dynamic pricing for perishable foods with quality and quantity deteriorating simultaneously under reference price effects. International Journal of Systems Science: Operations & Logistics 6, 346-355. https://doi.org/10.1080/23302674.2018.1465618

Fan, S.; Li, J.; Xia, Y.; Tian, X.; Guo, Z.; Huang, W. (2019): Long-term evaluation of soluble solids content of apples with biological variability by using near-infrared spectroscopy and calibration transfer method. Postharvest Biology and Technology 151, 79-87. https://doi.org/10.1016/j.postharvbio.2019.02.001

Felix Instruments (2020): Food science instruments. https://www.felixinstruments.com/food-science-instruments/portable-nir-analyzers/f-750-produce-quality-meter/, accessed on 21 June 2020

Giovenzana, V.; Beghi, R.; Buratti, S.; Civelli, R.; Guidetti, R. (2014): Monitoring of fresh-cut Valerianella locusta Laterr. shelf life by electronic nose and VIS-NIR spectroscopy. Talanta 120, 368-375. https://doi.org/10.1016/j.talanta.2013.12.014

Goisser, S.; Fernandes, M.; Ulrichs, C.; Mempel, H. (2018): Non-destructive measurement method for a fast quality evaluation of fruit and vegetables by using food-scanner. DGG-Proceedings 8, 1-5. https://doi.org/10.5288/DGG-PR-SG-2018

Goisser, S.; Mempel, H.; Bitsch, V. (2020): Food-Scanners as a Radical Innovation in German Fresh Produce Supply Chains. 101 - 116 Pages / International Journal on Food System Dynamics, Vol 11, No 2 (2020) / International Journal on Food System Dynamics, Vol 11, No 2. https://doi.org/10.18461/IJFSD.V11I2.43

Guthrie, J.A.; Walsh, K.B.; Reid, D.J.; Liebenberg, C.J. (2005): Assessment of internal quality attributes of mandarin fruit. 1. NIR calibration model development. Aust. J. Agric. Res. 56, 405. https://doi.org/10.1071/AR04257

Jannok, P.; Kamitani, Y.; Kawano, S. (2014): Development of a Common Calibration Model for Determining the Brix Value of Intact Apple, Pear and Persimmon Fruits by near Infrared Spectroscopy. Journal of Near Infrared Spectroscopy 22, 367-373. https://doi.org/10.1255/jnirs.1130

Kaur, H.; Künnemeyer, R.; McGlone, A. (2017): Comparison of hand-held near infrared spectrophotometers for fruit dry matter assessment. Journal of Near Infrared Spectroscopy 25, 267-277. https://doi.org/10.1177/0967033517725530

Li, J.; Xue, L.; Liu, M.H.; Lv, P.; Yan, L.Y. (2011): Determination of Moisture Content in Ginger Using PSO Combined with Vis/NIR. AMR 320, 563-568. https://doi.org/10.4028/www.scientific.net/AMR.320.563

McGlone, V.A.; Fraser, D.G.; Jordan, R.B.; Künnemeyer, R. (2003a): Internal quality assessment of mandarin fruit by vis/NIR spectroscopy. Journal of Near Infrared Spectroscopy 11, 323-332. https://doi.org/10.1255/jnirs.383

McGlone, V.A.; Jordan, R.B.; Seelye, R.; Clark, C.J. (2003b): Dry-matter–a better predictor of the post-storage soluble solids in apples? Postharvest Biology and Technology 28, 431-435. https://doi.org/10.1016/S0925-5214(02)00207-7

McGlone, V.A.; Kawano, S. (1998): Firmness, dry-matter and soluble-solids assessment of postharvest kiwifruit by NIR spectroscopy. Postharvest Biology and Technology 13, 131-141. https://doi.org/10.1016/S0925-5214(98)00007-6

Mehinagic, E.; Royer, G.; Bertrand, D.; Symoneaux, R.; Laurens, F.; Jourjon, F. (2003): Relationship between sensory analysis, penetrometry and visible-NIR spectroscopy of apples belonging to different cultivars. Food Quality and Preference 14, 473-484. https://doi.org/10.1016/S0950-3293(03)00012-0

Møller, S.M.; Travers, S.; Bertram, H.C.; Bertelsen, M.G. (2013): Prediction of postharvest dry matter, soluble solids content, firmness and acidity in apples (cv. Elshof) using NMR and NIR spectroscopy: a comparative study. Eur Food Res Technol 237, 1021-1024. https://doi.org/10.1007/s00217-013-2087-6

Ncama, K.; Magwaza, L.S.; Poblete-Echeverría, C.A.; Nieuwoudt, H.H.; Tesfay, S.Z.; Mditshwa, A. (2018): On-tree indexing of 'Hass' avocado fruit by non-destructive assessment of pulp dry matter and oil content. Biosystems Engineering 174, 41-49. https://doi.org/10.1016/j.biosystemseng.2018.06.011

OECD (2018): OECD fruit and vegetable scheme. Guidelines on objective tests to determine quality of fruit and vegetables, dry and dried produce

Olarewaju, O.O.; Bertling, I.; Magwaza, L.S. (2016): Non-destructive evaluation of avocado fruit maturity using near infrared spectroscopy and PLS regression models. Scientia Horticulturae 199, 229-236. https://doi.org/10.1016/j.scienta.2015.12.047

Parpinello, G.P.; Nunziatini, G.; Rombolà, A.D.; Gottardi, F.; Versari, A. (2013): Relationship between sensory and NIR spectroscopy in consumer preference of table grape (cv Italia). Postharvest Biology and Technology 83, 47-53. https://doi.org/10.1016/j.postharvbio.2013.03.013

Pasquini, C. (2003): Near Infrared Spectroscopy: Fundamentals, Practical Aspects and Analytical Applications. Journal of the Brazilian Chemical Society 14, 198-219. https://doi.org/10.1590/S0103-50532003000200006

Pöpping, B.; Bourdichon, F. (2018): Consumer food testing devices: threat or opportunity? New Food 21, 30-33

Rateni, G.; Dario, P.; Cavallo, F. (2017): Smartphone-Based Food Diagnostic Technologies: A Review. Sensors (Basel, Switzerland) 17. https://doi.org/10.3390/s17061453

Sánchez, M.-T.; Pérez-Marín, D.; Flores-Rojas, K.; Guerrero, J.-E.; Garrido-Varo, A. (2009): Use of near-infrared reflectance spectroscopy for shelf-life discrimination of green asparagus stored in a cool room under controlled atmosphere. Talanta 78, 530-536. https://doi.org/10.1016/j.talanta.2008.12.004

Santos Neto, J.P. dos; Assis, M.W.D. de; Casagrande, I.P.; Cunha Júnior, L.C.; Almeida Teixeira, G.H. de (2017): Determination of 'Palmer' mango maturity indices using portable near infrared (VIS-NIR) spectrometer. Postharvest Biology and Technology, 130, 75-80. https://doi.org/10.1016/j.postharvbio.2017.03.009

Schmilovitch, Z.; Mizrach, A.; Hoffman, A.; Egozi, H.; Fuchs, Y. (2000): Determination of mango physiological indices by near-infrared spectrometry. Postharvest Biology and Technology 19, 245-252. https://doi.org/10.1016/s0925-5214(00)00102-2.

Spectral Engines (2020): Food Scanner. The world's smartest, fastest and easiest way to measure food content. https://www.spectralengines.com/products/nirone-scanner/foodscanner, accessed on 21 June 2020.

Sunforest (2020): Portable Nondestructive Fruit Quality Meter. http://sunforest.kr/index.php?sm_idx=eng, accessed on 3 August 2020

Tellspec (2020): Empowering a Healthier World. with Real-Time Analysis Using Portable Low-Cost Sensors. https://tellspec.com/, accessed on 21 June 2020

UNECE, 2019: Fresh Fruit and Vegetables - Standards. http://www.unece.org/trade/agr/standard/fresh/ffv-standardse.html, accessed on 21 June 2020

Wang, H.; Peng, J.; Xie, C.; Bao, Y.; He, Y. (2015): Fruit Quality Evaluation Using Spectroscopy Technology: A Review. Sensors 15, 11889-11927. https://doi.org/10.3390/s150511889

Wedding, B.B.; Wright, C.; Grauf, S.; White, R.D.; Tilse, B.; Gadek, P. (2013): Effects of seasonal variability on FT-NIR prediction of dry matter content for whole Hass avocado fruit. Postharvest Biology and Technology 75, 9-16. https://doi.org/10.1016/j.postharvbio.2012.04.016

Authors

M.Sc. Simon Goisser, Research assistant at University of Applied Sciences Weihenstephan-Triesdorf, Department of Greenhouse Technology and Quality Management, Am Staudengarten 10, 85354 Freising, Germany

M.Sc. Sabine Wittmann, Research assistant at University of Applied Sciences Weihenstephan-Triesdorf, Department of Greenhouse Technology and Quality Management, Am Staudengarten 10, 85354 Freising, Germany

Prof. Dr. Heike Mempel, Professor at University of Applied Sciences Weihenstephan-Triesdorf and head of the department of Greenhouse Technology and Quality Management, Am Staudengarten 10, 85354 Freising, Germany. E-Mail: heike.mempel@hswt.de

Acknowledgements

The research of this publication was financed by the QS Science Funds in Fruit, Vegetables and Potatoes. The preparatory work for the study presented here was carried out within the framework of the alliance "Wir retten Lebensmittel" [We Save Foodstuffs], which was funded by the Bavarian Ministry of Food, Agriculture and Forestry. The authors address special thanks to STEP Systems GmbH from Nuremberg, Germany for their loan of two food-scanner devices during the course of these studies.

9 Conclusion and future prospects

9.1 Corroboration of hypotheses

This doctoral thesis examined the potential of portable and miniaturized NIR spectrometers as measuring instruments for non-destructive quality assessment of fruit and vegetables along the FSC. The basis for this work was laid by conducting a qualitative study with actors operating at different levels along the FSC and thereby demonstrating the potential of food-scanners as radical innovation for the fresh produce industry. Based on these findings, concrete measurements and experiments were carried out to determine the performance and practical suitability of these devices. By using tomato as model fruit, this research showed that several important quality parameters can be determined and predicted with a high degree of accuracy using portable and commercially available food-scanners. The transfer of the methods applied to tomato to other fruit and vegetable species also showed good correlations and illustrated the wide range of applications of these measuring instruments. These results provided new and significant contributions to scientific knowledge about miniaturized NIR spectrometers and possible applications along the FSC. During the course of a study at the department of incoming goods control, which has been described in Chapter 8, the suitability and applicability of these devices in current quality control processes could be demonstrated. In addition, all formulated hypotheses were adequately tested and confirmed within the framework of this doctoral thesis.

The first hypothesis of this thesis addressed the issue that actors along the FSC see great potential for the application of food-scanners. A qualitative study in Chapter 3 confirmed this hypothesis and identified food-scanners as a radical innovation for the German FSC. The results showed that at the moment, there are different practices of quality control for fresh produce in place along the FSC, and that those practices vary between production, wholesale and retail companies. The determination of quality is first and foremost very subjective and in many cases reduced to the examination of external appearance, especially at production and retail level. As for production cooperatives and wholesalers, compliance with established protocols is legally required to ensure quality and conformity of marketable produce with existing standards. In this context, actors along the FSC deemed portable and miniaturized food-scanners potential tools for a fast, non-destructive, cost-effective and objective quality measurement. These instruments could therefore help to overcome existing discrepancies in quality assessment along the FSC by introducing and providing a uniform measurement method. Furthermore, these devices could serve as decision-support tools and help untrained personnel, who do not yet have years of experience in quality control, to familiarise themselves more quickly with important quality attributes of fresh produce. The potential of these miniaturized spectrometers to predict various parameters non-destructively by means of a single spectrum is also specifically addressed in the current OECD guidelines on objective tests, enabling the development of "multidimensional predictors of consumer acceptance" (OECD 2018). In addition to the use of these devices in practice-oriented research projects such as the research projects on which this thesis is based and further German research projects BigApple (Biegert 2018) and Melon (KOB 2020), there are also reports describing their introduction to producer cooperatives such as the Australian Mango Industry Association (Edwards 2016) and the The Mexican Association of Avocado Producers and Exporting Packers (Henríquez 2019). Personal discussions with stakeholders at different levels of the fruit and vegetable supply chain in the course of the work on this dissertation as well as first cooperations with German fruit trading companies (see Chapter 8) further illustrate the great interest of the industry as well as the potential for practical application of food-scanners.

The second hypothesis stated that there are different areas of application for food-scanners along the FSC. Like the first hypothesis, this was confirmed primarily by the qualitative study in Chapter 3. As stated by actors in the conducted interviews, food-scanners can be used on a variety of fruit and on various levels of the FSC, e.g., for monitoring and determining fruit ripeness in production. Additionally, these instruments can be used at downstream levels of the FSC, e.g., production cooperatives and wholesalers, for quality validation during incoming goods control. At retail level, food-scanners can help to advertise and promote fruit quality with respect to taste or valuable ingredients, especially to health-conscious end-consumers. Overall, a high potential was identified for these portable and miniaturized NIR spectrometers in the documentation and profiling of fruit and vegetable quality along the entire supply chain. The experiment carried out for shelf-life modelling of tomato (see Chapter 5) also shows that food-scanners have a certain potential for estimating shelf life, especially when NIR measurements are combined with models that predict changes in fruit properties due to variances in ambient temperature and humidity. In turn, this could enable targeted decisions, and produce with longer shelf life could be kept in storage whereas produce with shorter shelf life could be forwarded faster. The use of food-scanners could therefore be a first step towards a change in distribution behavior, which consequently could lead to a reduction of food loss along the FSC. As illustrated by Hertog et al. (2014), the shift from a classical first-in-first-out (FIFO) towards a first-expired-first-out (FEFO) approach within the perishable supply chain is feasible by combining systems for monitoring the real-time conditions of the supply chain with specific algorithms for supply chain optimization. In addition, a recent study demonstrated the possibility of using NIR in on-line applications for the determination of shelf life in strawberries (Shen et al. 2018). Also, promising results were obtained for combining NIRS with electronic nose techniques for forecasting the days before decay of peach fruit (Huang et al. 2017). In a current research project, this idea is being developed further based on the findings of this thesis and a platform for optimizing the food supply chain using artificial intelligence and NIR spectra is being built (FreshAnalytics 2020).

Hypothesis three claimed that prediction models derived from food-scanner spectra show comparable results to those computed with benchtop NIR instruments. This was confirmed in an experiment which compared the performance of prediction models of important tomato quality parameters (sugar content, dry matter, firmness) developed with the SCiO™ and a laboratory NIR spectrometer (see Chapter 4). The independent multivariate analysis software The Unscrambler® was used for building of prediction models. The spectral data collected from both devices obtained high correlations for all three quality parameters. The direct comparison of the correlation coefficient r^2_{CV} and the root mean square error $RMSE_{CV}$ indicates a slightly higher prediction model accuracy for models built with spectra from the benchtop device. These differences are also comparable to the differences between the three portable devices SCiO™, F-750 and H-100F, as shown in chapter 5. Nevertheless, the models of the SCiO™ show high accuracy and can therefore be deemed suitable for predictions in practice. Findings from other publications, which compared the performance of portable and miniaturized spectrometers with laboratory instruments for a variety of produce such as acerola (Malegori et al. 2017), apples (Kaur et al. 2017; Khatiwada et al. 2016; Schmutzler and Huck 2016) and kiwifruit and stonefruit (Kaur et al. 2017) underline the results of this work. Displaying approximately the same model accuracy as a benchtop instrument, portable and miniaturized devices have clear advantages in practical applications, as their portability and size make variable use along the value chain easier to implement.

The fourth hypothesis of this thesis claimed that commercially available food-scanners achieve similar prediction model accuracy of selected fruit quality parameters. The studies conducted for the prediction of various tomato quality paramters (see Chapter 5) and the assessment of tomato lycopin content (see Chapter 7) were able to confirm this hypothesis. In addition, points of reference could be generated with regard to the accuracy of the models, in terms of the coefficient of determination r^2_{CV} and root meas square error (RMSE), with which the respective parameters can be predicted non-destructively. The NIR spectra of the three instruments used in these studies, SCiO™, F-750 and H-100F, achieved high correlations ($r^2_{CV} > 0.89$) for the quality attributes firmness, dry matter, sugar content, the chromaticity values L*, a*, h° and the secondary plant metabolite lycopene. These results are in line with reports about food-scanner performance cited in the scientific literature, e.g. the prediction of sugar content in pear (Choi et al. 2017) as well as the detection of dry matter content in kiwifruit and stonefruit (Kaur et al. 2017). It should also be noted that the results for food-scanners obtained in this thesis are in range with findings derived with laboratory NIR instruments, e.g. for the prediction of lycopene content in intact tomatoes (Clément et al. 2008; Li et al. 2017b), and even surpassed results for the prediction of lycopene content of processed tomato samples such as purée and ketchup

(Baranska et al. 2006; Szuvandzsiev et al. 2014). Furthermore, the results showed significantly lower accuracies (r^2_{CV} < 0.75) for the prediction of the quality parameters acidity, sugar-acid ratio and chromaticity values b* and C* with all three instruments. The difficulty of predicting acidity in tomatoes, even using laboratory NIR instruments, has been outlined in previous research (Flores et al. 2009; Oliveira et al. 2014). The results achieved in these studies are comparable to the prediction accuracies obtained in this thesis. As described by Oliveira et al. (2014), tomatoes are characterized by a low concentration of acidity and a heterogeneous composition of sugar content and acidity. Furthermore, tomatoes are divided into different compartments, so-called loculi, so they show no homogeneous composition of fruit. Each loculi is encompassed by a wall of fruit flesh, and this structure can cause interference with NIR radiation and be the reason for the mediocre results for acidity prediction in intact tomato. Further discrepancies between the predictions of the respective devices, in particular with regard to colour values, can be attributed to differences in wavelength ranges. Whereas the spectrometer of the F-750 operates in the bandwidth of 477 – 1059 nm and thereby covers almost the full visible spectrum, the H-100F utilizes a significantly smaller spectral range from 650 – 950 nm, which contains the red region of the visible spectrum. In contrast, the SCiO™ operates in a wavelength range from 740 – 1070 nm, which only covers a small range of the far-red end of the visible spectrum. Other differences within the technical specifications (e.g. spectral resolution, arrangement of light source and detector) also influence prediction accuracy and can lead to deviations between the three devices.

In general, the experiments conducted during the course of this thesis have shown that the three devices exhibit a comparable accuracy for important quality attributes of fruit, especially when very high correlations were obtained. The increasing importance of these portable measuring devices in the food sector is underlined by further studies, which investigate the application of food-scanners for the detection of total antioxidant capacity in gluten-free grains (Wiedemair and Huck 2018), the prediction of total protein and intact casein in cheddar cheese (Ma et al. 2019) and the monitoring of dry fermented sausages (González-Mohino et al. 2020). Furthermore, personal communications with several start-up companies, which develop portable and miniature NIR devices, such as trinamix GmbH (2020) and Senorics (2020), revealed that the agricultural and horticultural sector is an area of great interest for future applications.

Hypothesis five argued that software solutions provided by some food-scanner manufacturers achieve prediction model performance comparable to state-of-the-art multivariate analysis software. This hypothesis was confirmed in a comparative study (see Chapter 5). The direct comparison of prediction models for different quality parameters of tomato, which were created using SCiO™ Lab, Felix Instruments Model Builder and The Unscrambler®, obtained similar predictive performance, especially for models of high accuracy (r^2_{CV} > 0.90). With regard to the practical application of food-scanners, these models with high predictive accuracy are particularly important as they enable reliable measurements along the FSC. In summary, by confirming hypotheses three and four, this work was able to show that commercially available food-scanners as well as the software solutions provided are suitable tools for creating feasible non-destructive prediction models for various fruit quality traits. As a result, actors along the FSC can use these devices independently and do not need to rely on additional professional and expensive software for data analysis and model building. Nevertheless, it should be noted that the H-100F device evaluated in this thesis is not equipped with a software for building of own models. Communications with the device manufacturer (Hwang 10/15/2018) as well as a previous publication (Choi et al. 2017) revealed that such a software, called Sunforest PLSR Builder™, was distributed in the past, but was abandoned because it was not deemed necessary. The results presented by Choi et al. (2017) indicate that the software displays a performance comparable to the two software solutions evaluated in this thesis. Consequently, the independence of the H-100 series distributed by Sunforest could be improved by supplying users with this model builder software, omitting the need for additional software for multivariate analysis such as The Unscrambler®.

Hypothesis six stated that currently available devices differ in practicability and applicability with respect to the requirements of the fruit and vegetable supply chain. The detailed analysis of three commercially available food-scanners in Chapter 5 confirmed this hypothesis. Based on the findings of the qualitative study in Chapter 3, the devices were examined for a variety of criteria with regard to the independence of the devices, data management and practical handling. As the analysis illustrates, these three commercially available food-scanners meet the requirements expressed by actors along the value chain to varying degrees.

The F-750, which has been specifically developed for the measurement of fruit and vegetables, can be used as a stand-alone unit, which is characterized by a certain robustness due to the sturdy casing. In addition, the access to spectra as well as the software for building own prediction models allows easy handling and is a good prerequisite for the integration of measured values into existing systems of information technology. High battery durability and fast measurement speed compared to destructive measurements (see Chapter 8) are also favorable.

In a similar manner, the H-100F has been specially designed for fruit quality measurement and can be used as a stand-alone device which requires no additional equipment or internet access during operation. The durability of battery life and device robustness is comparable to the F-750, however the H-100F device only weighs half as much and performs measurements in a third of the F-750's time. In contrast to the F-750, the software provided for the H-100F does not allow the user to build own prediction models. As described earlier, such a software existed and has been provided to customers in the past, but was discontinued. In order to build and install own models on the device, users currently rely on additional software such as The Unscrambler®. As of now, the integration of the scanner into existing systems of information technology is therefore limited to the reading of measurement results.

In contrast, the miniaturized SCiO™ requires an additional device such as a tablet or smartphone as well as internet access via WiFi or mobile data to perform the scanning process. In general, this is not necessarily a disadvantage, as some companies already use tablets for data acquisition in incoming goods control. However, according to some supply chain actors, the small dimension of the SCiO™ could be a drawback, as such a small device could get lost more easily between pallets and produce during quality control. Due to the small dimension, a comparatively small battery is implemented, which in turn leads to a reduction in battery life compared to the F-750 and H-100F. However, the device benefits from the fast measuring speed. A major drawback is the difficult access to raw spectra, particularly in the context of the recent introduction of an annual fee for the use of the system. On the one hand, this makes implementation in existing systems of information technology more difficult. On the other hand, users become dependent on the sensor manufacturer, as independent data management and model creation is not possible. Termination of the licence would mean a loss of all gathered data and prediction models.

In summary, the F-750 and H-100 meet almost all necessary criteria. Both devices have been specifically developed for the measurement of fruit and their performance with regard to prediction accuracy of fruit quality traits has been the scope of many scientific studies in recent years (Choi et al. 2017; Kaur et al. 2017; Ncama et al. 2018; Sarkar et al. 2020; Scalisi and O'Connell 2020; Toivonen et al. 2017). The results of this thesis indicate that an improvement of the measurement speed of the F-750, while maintaining the current accuracy, could further improve the already reduced measurement effort compared to destructive measurements. As for the H-100F, a reintroduction of the model building software in combination with instructions for use could improve user friendliness and independent use by actors along the supply chain. In contrast, the characteristics of the SCiO™ deviate in many aspects from the requirements of the FSC. Especially the need for a licence to use the SCiO™ makes this system questionable with regard to future use along the FSC.

In addition to these three devices, which have been studied in great detail in this thesis, other portable and miniaturized NIR devices can be used to measure fruit quality and have been evaluated for this purpose in the literature (see Table 4).

As described in several studies, the DLP® NIRscan™ Nano sensor from Texas Instruments is oftentimes used in elaborate experimental setups for the detection of fruit quality (Al-Sanabani et al. 2019; Amirul et al. 2020; Mishra et al. 2021a; Sheng et al. 2019; Yuan et al. 2020), which involve 3D-printed cases and evaluation of spectra using additional software for multivariate analysis. This sensor can therefore not simply be used for measurements and own modelling of fruit quality parameters along the FSC. However, the DLP® NIRscan™ Nano is the sensor used by Tellspec® in its measurement device, and spectra can be evaluated using external software such as MATALB and The Unscrambler® (Marques and Freitas 2020) or the software provided by Tellspec® itself for machine learning model development (Santos et al. 2020). Since this data collection and management software is subject to an annual subscription fee (Tellspec Inc. 2020b), the implementation for utilization along the FSC is comparable to the SCiO™ and questionable for the future.

A literature research did not yield any reports for the NIRONE Sensors (Spectral Engines Oy) and the NeoSpectra-Scanner (Si-Ware Systems) for applications with respect to fruit quality detection. However, initial experiments regarding the identification of allergens in powdered food (Rady et al. 2020) and the monitoring of sugar content in breakfast cereals (Aykas et al. 2020) indicate the high predictive capability of these sensors. Since no concrete applications in the fresh produce sector are currently known and the possibility of model creation and user-friendliness of device handling in practice are unclear, the application along the FSC seems not viable at the moment.

The K-BA100R from the Kubota Coperation is advertised as a portable and non-destructive fruit analyzer (Focus Technology Co. 2020) and the results of various studies indicate a high predictive accuracy (Hu et al. 2019; Liu et al. 2015; Qi et al. 2017; Uwadaira et al. 2018). However, these studies used raw spectra and performed evaluations with external software such as MATLAB, R and The Unscrambler®. From the available literature and product information, it is not clear to what extent an adaptation of the existing models by users is possible. For this reason, no conclusive assessment of this device can be made on the implementation and use along the FSC.

The MicroNIR from VIAVI Solutions is designed for "customer needs across multiple industries" (VIAVI Solutions Inc. 2020). Besided quality assessment in the field of fruit and vegetables (Malegori et al. 2017; Pu et al. 2018; Sánchez et al. 2020), applications for this device include monitoring of food quality such as dry fermented sausage (González-Mohino et al. 2020) and classification of palm oil adulteration (Basri et al. 2017) and extend to other industries such as plastic waste sorting (Rani et al. 2019), pharmaceutical raw material identification (Sun et al. 2016) and detection of explosives (Risoluti et al. 2018). According to the distributers website, the MicroNIR is equipped with a software suite for data collection, calibration and model development. However it seems that a certain knowledge in chemometrics is necessary in order to build own prediction models. As no predesigned models for determining fruit quality are known, a definitive estimate for the utilization along the FSC is difficult.

According to the manufacturers, other devices used in the literature for measuring the quality of fruit, such as the MicroPhazir, Phazir 2400 and Phazir 1018, were developed primarily for identifying pharmaceutical raw materials (Thermo Fisher Scientific 2020) and sensing of material (Polychromix 2007). These devices are portable but significantly larger than the rest of the instruments discussed in this paper. Due to the lack of information on the possibility of building and utilizing own prediction models, the use of these devices for practical measurement of fruit quality along the FSC does not seem very realistic.

Hypothesis seven argued that food-scanners can be used for the prediction of various quality-related parameters on a wide range of fruit and vegetables. This hypothesis was confirmed in Chapter 8, where besides tomato a variety of other fruit was examined for the correlation of important quality parameters using NIR spectra of the F-750, H-100F and SCiO™. As demonstrated, important quality traits such as dry matter and sugar content of various produce can be predicted with good to high accuracy using commercial food-scanners. Additionally, the analysis of the relative water content of ginger obtained accurate predictive capability. Further tests on grapes and tomatoes using the example of sugar content also showed that non-destructive prediction is possible not only on bare fruit, but also through existing packaging foil. A comparison with results from the literature shows high consistency for the performance of these handheld and miniature devices, such as the determination of the dry matter content of apple (Zhang et al. 2019) and avocado (Subedi and Walsh 2020) as well as the detection of sugar content in table grape (Donis-González et al. 2020) and tangerines (Santos et al. 2020). These results illustrate the variety of possible applications along the FSC, starting with the multitude of possible products, the measurement of loose and packaged goods and ranging to applications at different stages along the value chain. In a first experimental approach, the possibility of estimating the shelf-life of tomatoes using food-scanners was also investigated (see Chapter 6). However, further research is necessary in order to make precise statements. In addition, other influencing parameters, such as air humidity and air circulation, must be taken into account in the underlying mathematical models.

9.2 Additional findings and recommendations

The investigations and experiments carried out in the course of this work also provided insights into the application and requirements for implementation of food-scanners along the FSC that went beyond the original hypotheses (Figure 10).

In the course of the qualitative survey of actors along the FSC in Chapter 3, various requirements were identified that food-scanners must fulfil in order to be of practical use. Besides device-specific requirements, the interviewed supply chain actors also highlighted the importance of reports about food-scanner in relevant media, such as newspapers and magazines, in order to gain trust with respect to the reliability and accuracy of these devices for fruit quality predictions. In order to gain broader attention and acceptance, results should not only be published in scientific journals, but also specialized magazines for practitioners. In this context, own experiences could already be made in the course of this doctoral thesis: The publication of the results of the research project on the website of the QS Science Funds have generated numerous positive reactions from practitioners and led to targeted enquiries for practical applications. Therefore, the integration into structures and platforms that are already utilized by many users offers a good way to inform, awaken interest and establish contact. Additional measures such as practical demonstrations, free trials or professional development courses have been highlighted in the literature for promoting innovations (Läpple et al. 2015; Pierpaoli et al. 2013) and could be suitable actions by manufacturers for introducing food-scanners to the FSC.

In the context of different areas of application along the FSC, the interviewees of the qualitative study emphasised an essential point: In order to enable long-term utilization of food-scanners, prediction models from different devices have to be transferable along the supply chain to guarantee consistent and accurate prediction results at different levels. Unfortunately, the transferability of models between several devices, as stated by supply chain actors, could not be verified in the course of this work for any of the three food-scanners. Due to the importance of this topic in numerous research areas of spectroscopy, studies on model transfer between devices have already been conducted. Initial findings indicate a good transferability of calibration models, especially when models are transferred from benchtop NIR spectrometers, used as master instruments, to portable and handheld devices (Bin et al. 2017; Yang et al. 2019). In order to ensure the cross-device use of accurate prediction models and therefore guarantee reliable measurements along the supply chain, this topic has to be further explored for food-scanners and their respective characteristics (wavelength range, spectral resolution, background noise, variation in hardware).

As a conclusion, future work regarding the application of food-scanners for fruit and vegetables should not only focus on device performance, but also consider applicability and practicalness in daily quality control processes. A Multi-Actor Approach (MAA), as described by Roussaki et al. (2019) for promoting the Internet of Things employment in Agriculture, could be a suitable solution for the implementation of food-scanners in the FSC. Such an approach could help to identify and address the most important needs of the various stakeholders with respect to implementation and utilizsation of food-scanners and could also help to facilitate the interoperability and standardization of spectra and data provided by various different device manufacturers.

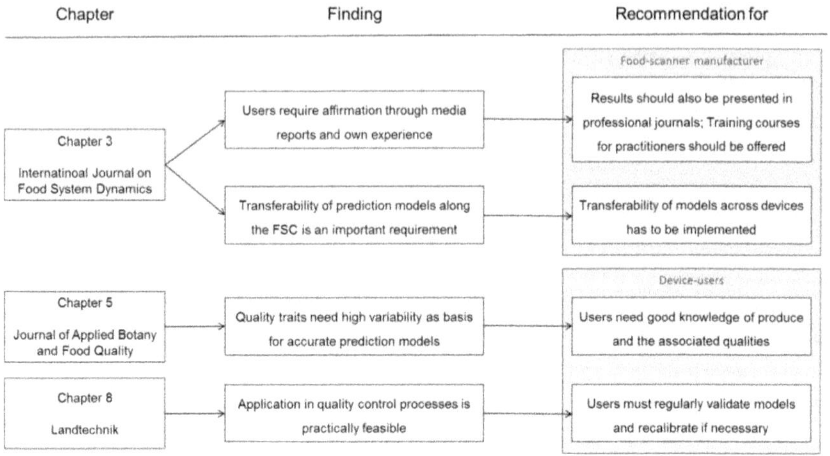

Figure 10: Additional findings and resulting recommendations for food-scanner manufacturers and device-users.

The study presented in Chapter 5 used two different experimental approaches in order to generate high variability of tomato fruit quality parameters. The results indicated that for the majority of quality traits a larger range of reference values yielded a higher accuracy of NIR prediction models. With regard to the development of new prediction models by device-users in day-to-day practice of quality control, this information is crucial and must be taken into account. Robustness and universal application is an ultimate aim for any calibration model (Zhang et al. 2017). Additionally, a standardized approach to calibration and validation is particularly important to develop routine. Currently, there is no such template for NIR utilization within the fruit and vegetable sector, although the potential of its application is explicitly mentioned in the OECD guidelines on objective tests (OECD 2018). Examples from other industries, such as pharmaceuticals, illustrate that official guidelines for the practical application of NIR have already been developed by pan-European committees (European Medicines Agency 2012). An elaborated document on the standard routine in the fruit and vegetable sector could be included in the Codex Alimentarius as a guideline or Code of Practice (FAO/WHO 2020). However, due to the diversity of the products as well as the natural variability in the fruit and vegetable sector, there is a high challenge in formulating these standards, as calibration solutions or calibration standards cannot be used here as in other research areas. In case of fresh produce, variability and thus robustness of models could be generated by using different cultivars, geographical origins and orchard locations, maturity stages and harvesting seasons (Fan et al. 2019; Magwaza et al. 2014; Wedding et al. 2013). However, extensive knowledge about the quality of fresh produce and influencing factors is an important prerequisite. Simple step-by-step instructions, as already used by some sensor manufacturers (Felix Instruments 2020b), could be an effective starting point for the development of such a standard.

Using a concrete practical example in Chapter 8, the applicability and practicability of food-scanners in day-to-day incoming goods control could be demonstrated. However, the presented use-case also illustrated that food-scanners are not simple plug-and-play devices. Values predicted by food-scanners have to be validated on a regular basis, especially when prediction models are new and yet not as robust due to lack of variability. Furthermore, the example of green table grapes also showed that after a certain time and the integration of different origins, a robust prediction model could be created. In the following season, this model enabled good quality predictions for new external batches of green table grapes. Users must therefore regularly validate calibration models and if necessary recalibrate these models by adding samples of new seasons, new cultivars or new origins. Care must be taken of the effort it takes to recalibrate these models, since too high complexity of a new technology could constitute a barrier for the adoption and implementation of food-scanners (Soderlund et al. 2008). The implementation of routine standards, as described above, in combination with training sessions could be a solution to this problem.

In the course of this work, different approaches to adapting the business strategy of food-scanner manufacturers could be observed. At the beginning of this doctoral thesis, start-up companies such as Consumer Physics and Tellspec® distributed food-scanners specifically designed for end-consumers. As of November 2020, Tellspec® specifies its market target as B2B and offers a web application for data collection and management, which requires an annual subscription fee to allow access and utilization of own measurements (Tellspec Inc. 2020a). The company also offers several mobile applications which allow the measurement of several fruit quality parameters. However, these measurements are not free of charge but must be purchased in a bundle, therefore every scan for the prediction of fruit quality is billed. In a similar manner, the business orientation of Consumer Physics, the distributer of the SCiO™ has changed. As of now, the website of Consumer Physics does not indicate any orientation towards end-consumers (Consumer Physcis 2020). Instead, numerous well-known companies (e.g., Corteva agriscience, Syngenta, Cargill) are listed as customers and partners, with whom together business solutions for the in-field analysis of maize, grain and animal feed are investigated. As already addressed in the previous chapter, the software provided for collecting spectra and evaluating prediction models which was previously for free is now subject to an annual fee. Furthermore, Consumer Physics announced the launch of a smartphone with an integrated SCiO™ sensor in January 2017, the Changhong H2 (ZDNet 2017), which should enable end-consumers to analyse the properties of food, liquids, medicines, and even measure body fat. However, a commercial launch of this smartphone was not carried out. These two examples illustrate the reorientation of start-up companies, that had originally targeted concepts for end-consumers, towards B2B solutions. The reason for this transformation of the business model has not been publicly announced. Previous research however showed that there are several barriers to innovation adoption by end-consumers (Joachim et al. 2018; Mani and Chouk 2017). A possible reason could be that food-scanners aimed at end-consumers have displayed a lack of perceived usefulness and a perceived price that is too high. A deficit in these important innovation characteristics could have increased consumers resistance to these products (Mani and Chouk 2017), which subsequently led to a reorganisation of manufacturers business models towards B2B.

Felix Instruments, the manufacturer of the F-750 Produce Quality Meter, has not changed its business orientation during the last three years. From the beginning, the company's primary target group have been fruit producers and production cooperatives. Therefore, the F-750 was predominantly designed for the in-field use in large orchards. However, during course of this work, successor devices to the F-750 have been released, which are specially designed for individual fruit such as avocado, mango and kiwi. These instruments only allow the measurement of the respective fruit, and modification or building of own calibration models is not possible. Personal communications with the manufacturer revealed that these new devices have been specifically designed for target customers, which oftentimes are specialized on certain tpyes of fruit. In addition, these specialized devices are characterized by a significantly lower price, which is an important purchase argument for many producers. In order to lower the barrier to innovatin adoption, the manufacturer revalued the innovation characteristics of its food-scanner by reducing the price and improving the usefulness for target customers (Mani and Chouk 2017).

9.3 Future prospects

In addition to the applications described in this thesis for determining fruit quality along the FSC, other possible applications for food-scanners in the field of fruit and vegetables are feasible in the future (Figure 11). In chapter 6, the possibility of shelf life analysis was presented using a first simple model. Based on these findings, new models for the shelf life of fruit and vegetables can be developed, which combine the previous measurement data from storage, such as temperature and humidity, with the non-destructive NIR measurements. This could allow an even more accurate estimation of the actual shelf life along the FSC and in the long run help to reduce food losses for suitable products. For a better assessment of the kinetic reaction of fruit and vegetables along the cooling chain so-called digital twins were used in an initial research project on mango (Defraeye et al. 2019). These digital twins contain all the elements of the real-world counterpart and are connected to real-world processes by input from sensor data, therefore allowing an accurate and realistic simulation of all relevant processes and process kinetics (Defraeye et al. 2019). Combining the digital twin approach with NIR measurements could optimise the food supply chain and improve the freshness, quality and safety of food. As a result, a research project, which is based on the data obtained in this work, is developing an internet platform with the help of artificial intelligence to optimise the food supply chain using mathematical shelf life models, digital twins of fruit, and NIR measurements (FreshAnalytics 2020).

In this context it should be noted that for the future it must be clarified who ultimately owns the generated data, who stores it, who has access to it and who is responsible for data security. Currently, for example, it is not clear who owns the data sovereignty for cloud solutions provided, such as The Lab application from SCiO™'s Consumer Physics. As already described in the examples of SCiO™ and Tellspec®, the business model of some manufacturers has also changed. The sole benefit is no longer the sale of the devices, but of the associated platforms for data management and bundles of scans. The qualitative interviews made clear how important this still open question is for the fruit trade sector, especially from the point of view of not making internal company information accessible to competitors and becoming dependent on manufacturers and annual subscriptions. In this context a privacy-aware management of internal company data is required to avoid identification, profiling, tracking and further privacy-related threats as described by Ziegeldorf et al. (2014).

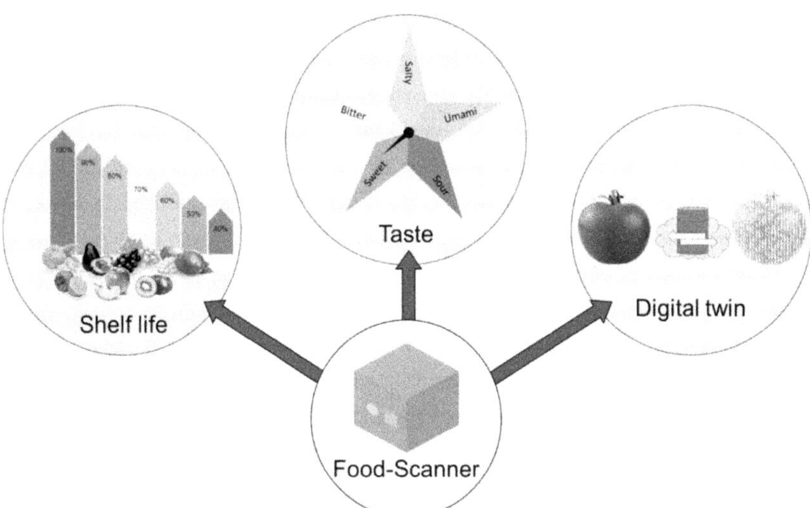

Figure 11: Potential future research areas for the application of NIR food-scanners.

Another potential area of application for food-scanners could be the non-destructive detection of taste characteristics. Although electronic noses are primarily used for this purpose, initial research work has also shown correlations of sensory taste properties of liquid foods such as wine and olive oil using NIR spectra (Cayuela et al. 2017; Inarejos-García et al. 2013). Preliminary studies on whole apples and table grapes have also shown a statistical correlation between NIR spectra and taste properties, and very good correlations between spectra and consumer preference for a particular taste class (Mehinagic et al. 2003; Parpinello et al. 2013). Based on these findings, the use of portable food-scanners for the non-destructive detection of taste would be another future field of research. In contrast to the work presented in this thesis, which used quantitative detection of fruit quality parameters, the characterization of taste via NIR food-scannes could make use of qualitative evaluations and clustering of taste profiles.

With regard to the practical use of these instruments along the supply chain, there are currently remaining uncertainties regarding the frequency and the intervals between validations and recalibrations of existing prediction models. Improvements within software and spectra processing could add significant value, e.g. the use of algorithms to recognise spectra already in the data pool. Therefore, fruits whose spectrum differs significantly from the previous data set could be selected for targeted reference measurement already during scanning. In this way, the variability of the collections can be specifically extended and costs and time can be saved by avoiding the measurement of unnecessary reference values.

9.4 Scientific and practical contribution

The demand on quality in the fruit and vegetable sector continues to increase beyond the legal minimum requirements. Measuring instruments that determine the internal quality of fruit quickly and non-destructively can be of considerable advantage in practice. This doctoral thesis focused on postharvest levels of the FSC and illustrates that food-scanners are suitable measuring devices for various parameters of fruit quality. The results presented in this paper not only promote new scientific knowledge regarding the measurement accuracy of these instruments, but also provide important insights with regard to their use in the practical work of day-to-day quality measurement.

This dissertation used a qualitative research approach to investigate the perception and assessment of actors along the FSC regarding the implementation and utilization of food-scanners for measuring fruit quality. Such an approach is suitable when not well-known or new research subjects are explored (Bitsch 2005). In the framework of innovation, the results obtained were able to show that food-scanners can be categorized as radical innovation for the German FSC. The conducted semi-structured interviews helped to illustrate the status quo of practical quality assessment along the German FSC, an area not well decribed in the literature. Furthermore, the results helped to identify important drivers and barriers for the implementation of food-scanners in daily processes of quality control. These results are of high importance not only for the scientific literature, but also for practitioners and sensor manufacturers, as they provide important information for the adaptation of food-scanners in the field of fruit and vegetables.

Based on these findings, this thesis focused on the performance of three commercially available hand-held and miniaturized food-scanners for the non-destructive prediction of important tomato quality parameters. In line with similar studies conducted at the time of writing, which also investigated these devices but other types of fruit such as apple (Kaur et al. 2017; Zhang et al. 2019), avocado (Li et al. 2018; Ncama et al. 2018), grapes (Momin et al. 2019) or mango (dos Santos Neto et al. 2018; Wokadala et al. 2020), the results presented here illustrate the high predictive capability of these instruments. Furthermore, the prediction accuracy of these commercially available devices for secondary plant metabolites could be demonstrated for the first time using the example of lycopene, an achievement that until then had only been possible using self-developed prototypes of miniaturized spectrometers (Sheng et al. 2019).

In addition to the solely scientific consideration of the prediction accuracy and based on the findings of the qualitative study, aspects that are of great importance for the use of these devices in everyday practice were taken into account. Therefore, an exemplary framework, which can help practitioners in evaluating food-scanner characteristics regarding usability and applicability, was designed. Additionally, a detailed comparison of software for spectra evaluation provided by the manufacturers of the SCiO™ and F-750 with state-of-the-art software for chemometrical analysis, which has previously only been performed for the SCiO™ to a limited extent in a single study (Li et al. 2018), showed a high degree of conformity. For use in practice, this means that no expensive and specialized evaluation software and the associated knowledge is required and device users can rely on software solutions provided by manufacturers.

Building on the results obtained on tomatoes the predictability of important fruit quality parameters of various types of fruit was examined. In addition, experiments on tomato and table grape showed that food-scanner measurements achieve good prediction results through packaging foil. For the practical application of food-scanner, this could bring an additional advantage, as it means that packaged produce does not have to be opened during quality control. The actual practical use of a commercial food-scanner in quality control in the incoming goods department confirmed the largely theoretical findings gathered up to that point. On the one hand, accurate prediction models could be created in a short time and used for further precise determination of the average quality of daily batches. On the other hand, the test showed that a considerable amount of time can be saved with the help of this non-destructive measuring method. Thus, in addition to the measurement accuracy of food-scanners, this doctoral thesis was able to demonstrate their concrete practical application along the FSC.

10 Declaration

I hereby declare that the work presented in this dissertation, titled

"Suitability of portable NIR sensors (food-scanners) for the determination of fruit quality along the supply chain using the example of tomatoes"

is an authentic record of my personal work.

I explicitly declare that I have marked all third-party sources used in this work, including those from the internet (tables, figures, graphics, etc.) as such. In particular, I confirm that both varbal statements and unmodified tables, figures, and the like as well as statements of other authors reproduced in my own words (paraphrase) or modified tables, figures and the like are marked as such.

Freising, March 27th , 2021

Simon Goisser

11 References

Abbott, Judith A. (1999): Quality measurement of fruits and vegetables. In Postharvest Biology and Technology 15 (3), pp. 207–225. DOI: 10.1016/S0925-5214(98)00086-6.

Acharya, U. K.; Subedi, P. P.; Walsh, K. B. (2017): Robustness of Tomato Quality Evaluation Using a Portable Vis-SWNIRS for Dry Matter and Colour. In International Journal of Analytical Chemistry 2017, p. 2863454. DOI: 10.1155/2017/2863454.

Al-Sanabani, Dheya Galal Abdullah; Solihin, Mahmud Iwan; Pui, Liew Phing; Astuti, Winda; Ang, Chun Kit; Hong, Lim Wei (2019): Development of non-destructive mango assessment using Handheld Spectroscopy and Machine Learning Regression. In Journal of Physics: Conference Series 1367, p. 12030. DOI: 10.1088/1742-6596/1367/1/012030.

Amirul, M. S.; Endut, R.; Rashidi, C. B. M.; Aljunid, S. A.; Ali, N.; Laili, M. H. et al. (2020): Estimation of Harumanis (Mangifera indica L.) Sweetness using Near-Infrared (NIR) Spectroscopy. In IOP Conference Series: Materials Science and Engineering 767, p. 12070. DOI: 10.1088/1757-899X/767/1/012070.

Amuah, Charles L. Y.; Teye, Ernest; Lamptey, Francis Padi; Nyandey, Kwasi; Opoku-Ansah, Jerry; Adueming, Peter Osei-Wusu (2019): Feasibility Study of the Use of Handheld NIR Spectrometer for Simultaneous Authentication and Quantification of Quality Parameters in Intact Pineapple Fruits. In Journal of Spectroscopy 2019, pp. 1–9. DOI: 10.1155/2019/5975461.

Anderson, N. T.; Walsh, K. B.; Subedi, P. P.; Hayes, C. H. (2020): Achieving robustness across season, location and cultivar for a NIRS model for intact mango fruit dry matter content. In Postharvest Biology and Technology 168, p. 111202. DOI: 10.1016/j.postharvbio.2020.111202.

Anderson, Nicholas T.; Subedi, Phul P.; Walsh, Kerry B. (2017): Manipulation of mango fruit dry matter content to improve eating quality. In Scientia Horticulturae 226, pp. 316–321. DOI: 10.1016/j.scienta.2017.09.001.

Ariana, Diwan P.; Lu, Renfu (2010): Evaluation of internal defect and surface color of whole pickles using hyperspectral imaging. In Journal of Food Engineering 96 (4), pp. 583–590. DOI: 10.1016/j.jfoodeng.2009.09.005.

Aykas, Didem Peren; Ball, Christopher; Menevseoglu, Ahmed; Rodriguez-Saona, Luis E. (2020): In Situ Monitoring of Sugar Content in Breakfast Cereals Using a Novel FT-NIR Spectrometer. In Applied Sciences 10 (24), p. 8774. DOI: 10.3390/app10248774.

Baca-Bocanegra, Berta; Hernández-Hierro, José Miguel; Nogales-Bueno, Julio; Heredia, Francisco José (2019): Feasibility study on the use of a portable micro near infrared spectroscopy device for the "in vineyard" screening of extractable polyphenols in red grape skins. In Talanta 192, pp. 353–359. DOI: 10.1016/j.talanta.2018.09.057.

Bai, Wenming; Yoshimura, Norio; Takayanagi, Masao (2014): Quantitative Analysis of Ingredients of Blueberry Fruits by near Infrared Spectroscopy. In Journal of Near Infrared Spectroscopy 22 (5), pp. 357–365. DOI: 10.1255/jnirs.1129.

Baranska, M.; Schütze, W.; Schulz, H. (2006): Determination of lycopene and beta-carotene content in tomato fruits and related products: Comparison of FT-Raman, ATR-IR, and NIR spectroscopy. In Analytical Chemistry 78 (24), pp. 8456–8461. DOI: 10.1021/ac061220j.

Basri, Katrul Nadia; Hussain, Mutia Nurulhusna; Bakar, Jamilah; Sharif, Zaiton; Khir, Mohd Fared Abdul; Zoolfakar, Ahmad Sabirin (2017): Classification and quantification of palm oil adulteration via portable NIR spectroscopy. In Spectrochimica Acta Part A: Molecular and Biomolecular Spectroscopy 173, pp. 335–342. DOI: 10.1016/j.saa.2016.09.028.

Batu, Ali (2004): Determination of acceptable firmness and colour values of tomatoes. In Journal of Food Engineering 61 (3), pp. 471–475. DOI: 10.1016/S0260-8774(03)00141-9.

Beć, Krzysztof B.; Grabska, Justyna; Huck, Christian W. (2020): Principles and Applications of Miniaturized Near-Infrared (NIR) Spectrometers. In Chemistry - A European Journal. DOI: 10.1002/chem.202002838.

Beghi, R.; Spinardi, A.; Bodria, L.; Mignani, I.; Guidetti, R. (2013): Apples Nutraceutic Properties Evaluation Through a Visible and Near-Infrared Portable System. In Food Bioprocess Technol 6 (9), pp. 2547–2554. DOI: 10.1007/s11947-012-0824-7.

Bertin, Nadia; Génard, Michel (2018): Tomato quality as influenced by preharvest factors. In Scientia Horticulturae 233, pp. 264–276. DOI: 10.1016/j.scienta.2018.01.056.

Biegert, Konni (2018): Mit BigApple zu weniger Lagerverlusten. In Poma, August 2018, pp. 18–21. Available online at https://inovel.de/file/POMA-August-2018-Mit-BigApple-zu-weniger-Lagerverlusten.pdf.

Bin, Jun; Li, Xin; Fan, Wei; Zhou, Ji-Heng; Wang, Cheng-Wei (2017): Calibration transfer of near-infrared spectroscopy by canonical correlation analysis coupled with wavelet transform. In The Analyst 142 (12), pp. 2229–2238. DOI: 10.1039/c7an00280g.

Birth, G. S.; Dull, G. G.; Renfroe, W. T.; Kays, S. J. (1985): Nondestructive Spectrophotometric Determination of Dry Matter in Onions. In Journal of the American Society for Horticultural Science 110 (2), pp. 297–303.

Bitsch, Vera (2005): Qualitative Research: A Grounded Theory Example and Evaluation Criteria. In Journal of Agribusiness 23 (1), pp. 75–91. DOI: 10.22004/ag.econ.59612.

Blakey, Robert J. (2016): Evaluation of avocado fruit maturity with a portable near-infrared spectrometer. In Postharvest Biology and Technology 121, pp. 101–105. DOI: 10.1016/j.postharvbio.2016.06.016.

Blanco-Díaz, María Teresa; Del Río-Celestino, Mercedes; Martínez-Valdivieso, Damián; Font, Rafael (2014): Use of visible and near-infrared spectroscopy for predicting antioxidant compounds in summer squash (Cucurbita pepo ssp pepo). In Food Chemistry 164, pp. 301–308. DOI: 10.1016/j.foodchem.2014.05.019.

Bokobza, L. (2002): Origin of Near-infrared Absorption Bands. In H. W. Siesler, Y. Ozaki, S. Kawata, H. M. Heise (Eds.): Near-Infrared Spectroscopy. Principles, Instruments, Applications. Weinheim: WILEY-VCH Verlag GmbH, pp. 11–42.

Brandt, Sára; Pék, Zoltán; Barna, Éva; Lugasi, Andrea; Helyes, Lajos (2006): Lycopene content and colour of ripening tomatoes as affected by environmental conditions. In Journal of the Science of Food and Agriculture 86 (4), pp. 568–572. DOI: 10.1002/jsfa.2390.

Brito, Anna Luiza Bizerra; Brito, Lívia Rodrigues; Honorato, Fernanda Araújo; Pontes, Márcio José Coelho; Pontes, Liliana Fátima Bezerra Lira (2013): Classification of cereal bars using near infrared spectroscopy and linear discriminant analysis. In Food Research International 51 (2), pp. 924–928. DOI: 10.1016/j.foodres.2013.02.014.

Brown, Steven D. (2016): The chemometrics revolution re-examined. In Journal of Chemometrics 31 (1), 1-23. DOI: 10.1002/cem.2856.

Buccheri, Marina; Grassi, Maurizio; Lovati, Fabio; Petriccione, Milena; Rega, Pietro; Scalzo, Roberto L.; Cattaneo, Tiziana M. P. (2019): Near infrared spectroscopy in the supply chain monitoring of Annurca apple. In Journal of Near Infrared Spectroscopy 27 (1), pp. 86–92. DOI: 10.1177/0967033518821829.

Camps, Cédric; Christen, Danilo (2009): On-tree follow-up of apricot fruit development using a hand-held NIR instrument. In Journal of Food, Agriculture & Environment 7 (2), pp. 394–400.

Casals, Joan; Rivera, Ana; Sabaté, Josep; Del Romero Castillo, Roser; Simó, Joan (2019): Cherry and Fresh Market Tomatoes: Differences in Chemical, Morphological, and Sensory Traits and Their Implications for Consumer Acceptance. In Agronomy 9 (1), p. 9. DOI: 10.3390/agronomy9010009.

Castrignanò, Annamaria; Buttafuoco, Gabriele; Malegori, Cristina; Genorini, Emiliano; Iorio, Raffaele; Stipic, Marija et al. (2019): Assessing the Feasibility of a Miniaturized Near-Infrared Spectrometer in Determining Quality Attributes of San Marzano Tomato. In Food Analytical Methods 12 (7), pp. 1497–1510. DOI: 10.1007/s12161-019-01475-x.

Cayuela, José Antonio; Puertas, Belén; Cantos-Villar, Emma (2017): Assessing wine sensory attributes using Vis/NIR. In European Food Research and Technology 243 (6), pp. 941–953. DOI: 10.1007/s00217-016-2807-9.

Chalmers, David J.; Rowan, Kingsley S. (1971): The Climacteric in Ripening Tomato Fruit. In Plant Physiology 48, pp. 235–240. DOI: 10.1104/pp.48.3.235.

Choi, Jin-Ho; Chen, Po-An; Lee, ByulHaNa; Yim, Sun-Hee; Kim, Myung-Su; Bae, Young-Seok et al. (2017): Portable, non-destructive tester integrating VIS/NIR reflectance spectroscopy for the detection of sugar content in Asian pears. In Scientia Horticulturae 220, pp. 147–153. DOI: 10.1016/j.scienta.2017.03.050.

Clark, C. J.; McGlone, V. A.; Jordan, R. B. (2003): Detection of Brownheart in 'Braeburn' apple by transmission NIR spectroscopy. In Postharvest Biology and Technology 28 (1), pp. 87–96. DOI: 10.1016/S0925-5214(02)00122-9.

Clément, Alain; Dorais, Martine; Vernon, Marcia (2008): Nondestructive measurement of fresh tomato lycopene content and other physicochemical characteristics using visible-NIR spectroscopy. In Journal of Agricultural and Food Chemistry 56 (21), pp. 9813–9818. DOI: 10.1021/jf801299r.

Consumer Physcis (2020): SCiO. Real-time decision making in the field with portable lab-grade analysis solutions. Available online at https://www.consumerphysics.com/, updated on 9/28/2020.

Cortés, V.; Blasco, J.; Aleixos, N.; Cubero, S.; Talens, P. (2019): Monitoring strategies for quality control of agricultural products using visible and near-infrared spectroscopy: A review. In Trends in Food Science & Technology 85, pp. 138–148. DOI: 10.1016/j.tifs.2019.01.015.

Costa, G.; Noferini, M.; Fiori, G.; Miserocchi, O.; Bregoli, A. M. (2002): NIRS Evaluation of Peach and Nectarine Fruit Quality in Pre- and Post-Harvest Conditions. In Acta Hortic. (592), pp. 593–599. DOI: 10.17660/ActaHortic.2002.592.81.

Cozzolino, Daniel (2009): Near infrared spectroscopy in natural products analysis. In Planta medica 75 (7), pp. 746–756. DOI: 10.1055/s-0028-1112220.

Cozzolino, Daniel (2020): The Sample, the Spectra and the Maths-The Critical Pillars in the Development of Robust and Sound Applications of Vibrational Spectroscopy. In Molecules 25 (16), p. 3674. DOI: 10.3390/molecules25163674.

Cuq, Sebastien; Lemetter, Valerie; Kleiber, Didier; Levasseur-Garcia, Cecile (2020): Assessing macro-element content in vine leaves and grape berries of vitis vinifera by using near-infrared spectroscopy and chemometrics. In International Journal of Environmental Analytical Chemistry 100 (10), pp. 1179–1195. DOI: 10.1080/03067319.2019.1648644.

Das, Anshuman J.; Wahi, Akshat; Kothari, Ishan; Raskar, Ramesh (2016): Ultra-portable, wireless smartphone spectrometer for rapid, non-destructive testing of fruit ripeness. In Scientific reports 6, p. 32504. DOI: 10.1038/srep32504.

Defraeye, Thijs; Tagliavini, Giorgia; Wu, Wentao; Prawiranto, Kevin; Schudel, Seraina; Assefa Kerisima, Mekdim et al. (2019): Digital twins probe into food cooling and biochemical quality changes for reducing losses in refrigerated supply chains. In Resources, Conservation and Recycling 149, pp. 778–794. DOI: 10.1016/j.resconrec.2019.06.002.

Dickens, Jason E. (2010): Overview of Process Analysis and PAT. In Katherine A. Bakeev (Ed.): Process Analytical Technology. Spectroscopic Tools and Implementation Strategies for the Chemical and Pharmaceutical Industries. Second Edition. West Sussex, United Kingdom: John Wiley & Sons, Ltd, pp. 1–15.

Donis-González, Irwin R.; Valero, Constantino; Momin, Md Abdul; Kaur, Amanjot; C. Slaughter, David (2020): Performance Evaluation of Two Commercially Available Portable Spectrometers to Non-Invasively Determine Table Grape and Peach Quality Attributes. In Agronomy 10 (1), p. 148. DOI: 10.3390/agronomy10010148.

Dos Santos, Cláudia A. Teixeira; Lopo, Miguel; Páscoa, Ricardo N. M. J.; Lopes, João A. (2013): A review on the applications of portable near-infrared spectrometers in the agro-food industry. In Applied spectroscopy 67 (11), pp. 1215–1233. DOI: 10.1366/13-07228.

Dos Santos Neto, João Paixão; Assis, Mateus Wagner Dantas de; Casagrande, Izabella Parkutz; Cunha Júnior, Luis Carlos; Almeida Teixeira, Gustavo Henrique de (2017): Determination of 'Palmer' mango maturity indices using portable near infrared (VIS-NIR) spectrometer. In Postharvest Biology and Technology 130, pp. 75–80. DOI: 10.1016/j.postharvbio.2017.03.009.

Dos Santos Neto, João Paixão; Leite, Gustavo Walace Pacheco; Da Oliveira, Gabriele Silva; Cunha Júnior, Luís Carlos; Gratão, Priscila Lupino; Morais, Camilo Lelis Medeiros de; Teixeira, Gustavo Henrique de Almeida (2018): Cold storage of 'Palmer' mangoes sorted based on dry matter content using portable near infrared (VIS-NIR) spectrometer. In Journal of Food Processing and Preservation 42 (6), e13644. DOI: 10.1111/jfpp.13644.

Edwards, Judith (2016): Australian mango season kicks off with help from the F-750 Produce Quality Meter! Available online at https://www.postharvest.biz/en/company/australian-mango-season-kicks-off-with-help-from-the-f-750-produce-quality-meter/_id:62863,seccion:news,noticia:77797/, checked on 12/16/2020.

Eisenstecken, Daniela; Stürz, Barbara; Robatscher, Peter; Lozano, Lidia; Zanella, Angelo; Oberhuber, Michael (2019): The potential of near infrared spectroscopy (NIRS) to trace apple origin: Study on different cultivars and orchard elevations. In Postharvest Biology and Technology 147, pp. 123–131. DOI: 10.1016/j.postharvbio.2018.08.019.

Entrenas, José-Antonio; Pérez-Marín, Dolores; Torres, Irina; Garrido-Varo, Ana; Sánchez, María-Teresa (2019): Safety and quality issues in summer squashes using handheld portable NIRS sensors for real-time decision making and for on-vine monitoring. In Journal of the Science of Food and Agriculture 99 (15), pp. 6768–6777. DOI: 10.1002/jsfa.9959.

Entrenas, José-Antonio; Pérez-Marín, Dolores; Torres, Irina; Garrido-Varo, Ana; Sánchez, María-Teresa (2020): Simultaneous detection of quality and safety in spinach plants using a new generation of NIRS sensors. In Postharvest Biology and Technology 160, p. 111026. DOI: 10.1016/j.postharvbio.2019.111026.

Escribano, S.; Biasi, W. V.; Lerud, R.; Slaughter, D. C.; Mitcham, E. J. (2017): Non-destructive prediction of soluble solids and dry matter content using NIR spectroscopy and its relationship with sensory quality in sweet cherries. In Postharvest Biology and Technology 128, pp. 112–120. DOI: 10.1016/j.postharvbio.2017.01.016.

European Medicines Agency (2012): Guideline on the use of Near Infrared Spectroscopy (NIRS) by the pharmaceutical industry and the data requirements for new submissions and variations. Available online at https://www.ema.europa.eu/en/documents/scientific-guideline/draft-guideline-use-near-infrared-spectroscopy-pharmaceutical-industry-data-requirements-new_en-0.pdf.

Fan, Shuxiang; Li, Jiangbo; Xia, Yu; Tian, Xi; Guo, Zhiming; Huang, Wenqian (2019): Long-term evaluation of soluble solids content of apples with biological variability by using near-infrared spectroscopy and calibration transfer method. In Postharvest Biology and Technology 151, pp. 79–87. DOI: 10.1016/j.postharvbio.2019.02.001.

Fan, Shuxiang; Wang, Qingyan; Tian, Xi; Yang, Guiyan; Xia, Yu; Li, Jiangbo; Huang, Wenqian (2020): Non-destructive evaluation of soluble solids content of apples using a developed portable Vis/NIR device. In Biosystems Engineering 193, pp. 138–148. DOI: 10.1016/j.biosystemseng.2020.02.017.

FAO/WHO (2020): Codex Alimentarius. International Food Standards. Edited by FAO/WHO. Available online at http://www.fao.org/fao-who-codexalimentarius/home/it/, checked on 12/22/2020.

Felix Instruments (2020a): F-750 Operation Manual. Edited by Felix Instruments. Camas, WA. Available online at https://www.felixinstruments.com/static/media/uploads/manuals/sharepoint_manual_f750.html, updated on 2/26/2020, checked on 12/11/2020.

Felix Instruments (2020b): Food Science Instruments. Portable NIR Analysis. Available online at https://www.felixinstruments.com/food-science-instruments/portable-nir-analyzers/, checked on 9/28/2020.

Ferrer-Gallego, Raúl; Hernández-Hierro, José Miguel; Rivas-Gonzalo, Julián C.; Escribano-Bailón, M. Teresa (2011): Determination of phenolic compounds of grape skins during ripening by NIR spectroscopy. In LWT - Food Science and Technology 44 (4), pp. 847–853. DOI: 10.1016/j.lwt.2010.12.001.

Flores, Katherine; Sánchez, María-Teresa; Pérez-Marín, Dolores; Guerrero, José-Emilio; Garrido-Varo, Ana (2009): Feasibility in NIRS instruments for predicting internal quality in intact tomato. In Journal of Food Engineering 91 (2), pp. 311–318. DOI: 10.1016/j.jfoodeng.2008.09.013.

Focus Technology Co. (2020): K-BAL00R Series Portable Fruit Nondestructive Analyzer. Available online at https://zjtop17.en.made-in-china.com/product/iXRmQSDlOfkW/China-K-BAL00R-Series-Portable-Fruit-Nondestructive-Analyzer.html, checked on 12/18/2020.

Fraser, Daniel D.; Künnemeyer, Rainer; McGlone, V. Andrew; Jordan, Robert B. (2001): Letter to the Editor. Near infra-red (NIR) light penetration into an apple. In Postharvest Biology and Technology 22 (3), pp. 191–195. DOI: 10.31436/ijes.v4i2.129.g78.

FreshAnalytics (2020): Eine Plattform zur KI Optimierung der Lebensmittellieferkette. Available online at https://www.freshanalytics.eu/, checked on 11/27/2020.

Fu, Xiaping; Ying, Yibin; Lu, Huishan; Xu, Huirong (2007): Comparison of diffuse reflectance and transmission mode of visible-near infrared spectroscopy for detecting brown heart of pear. In Journal of Food Engineering 83 (3), pp. 317–323. DOI: 10.1016/j.jfoodeng.2007.02.041.

Fu, Xiaping; Ying, Yibin; Xu, Huirong; Yu, Haiyan (2008): Support Vector Machines and Near Infrared Spectroscopy for Quantification of Vitamin C Content in Kiwifruit. An ASABE Meeting Presentation. In : 2008 ASABE Annual International Meeting, vol. 085204. 2008 Providence, Rhode Island, June 29 - July 2, 2008. St. Joseph, MI: American Society of Agricultural and Biological Engineers, pp. 1–9.

Gabriëls, Suzan H.E.J.; Mishra, Puneet; Mensink, Manon G.J.; Spoelstra, Patrick; Woltering, Ernst J. (2020): Non-destructive measurement of internal browning in mangoes using visible and near-infrared spectroscopy supported by artificial neural network analysis. In Postharvest Biology and Technology 166, p. 111206. DOI: 10.1016/j.postharvbio.2020.111206.

Geyer, Martin; Herold, Bernd; Zude, Manuela; Truppel, Ingo (2007): Non-Destructive Evaluation of Apple Fruit Maturity on the Tree. In Vegetable Crops Research Bulletin 66 (1). DOI: 10.2478/v10032-007-0018-4.

Goke, Alex; Serra, Sara; Musacchi, Stefano (2018): Postharvest Dry Matter and Soluble Solids Content Prediction in d'Anjou and Bartlett Pear Using Near-infrared Spectroscopy. In HortScience 53 (5), pp. 669–680. DOI: 10.21273/HORTSCI12843-17.

Goke, Alex; Serra, Sara; Musacchi, Stefano (2020): Manipulation of Fruit Dry Matter via Seasonal Pruning and Its Relationship to d'Anjou Pear Yield and Fruit Quality. In Agronomy 10 (6), p. 897. DOI: 10.3390/agronomy10060897.

González-Mohino, Alberto; Pérez-Palacios, Trinidad; Antequera, Teresa; Ruiz-Carrascal, Jorge; Olegario, Lary Souza; Grassi, Silvia (2020): Monitoring the Processing of Dry Fermented Sausages with a Portable NIRS Device. In Foods 9 (9), pp. 1–12. DOI: 10.3390/foods9091294.

GP Technical (2020): ASD Systems AgriSpec Compact Analyzer. Available online at https://gp-technical.com/product/asd-systems-agrispec-compact-nir-analyzer/, checked on 10/20/2020.

Greensill, Colin V.; Walsh, Kerry B. (2000): A remote acceptance probe and illumination configuration for spectral assessment of internal attributes of intact fruit. In Meas. Sci. Technol. 11 (12), pp. 1674–1684. DOI: 10.1088/0957-0233/11/12/304.

Guo, Ying; Ni, Yongnian; Kokot, Serge (2016): Evaluation of chemical components and properties of the jujube fruit using near infrared spectroscopy and chemometrics. In Spectrochimica acta. Part A, Molecular and biomolecular spectroscopy 153, pp. 79–86. DOI: 10.1016/j.saa.2015.08.006.

Henríquez, Priscila (2019): Technological advances are revolutionizing regional trade in Mexican avocados. IICA. Mexico City. Available online at https://www.iica.int/en/press/news/technological-advances-are-revolutionizing-regional-trade-mexican-avocados, checked on 12/16/2020.

Hernández-Hierro, José Miguel; Valverde, Juan; Villacreces, Salvador; Reilly, Kim; Gaffney, Michael; González-Miret, Maria Lourdes et al. (2012): Feasibility study on the use of visible-near-infrared spectroscopy for the screening of individual and total glucosinolate contents in broccoli. In Journal of Agricultural and Food Chemistry 60 (30), pp. 7352–7358. DOI: 10.1021/jf3018113.

Herschel, W. (1800): Investigation of the powers of the prismatic colours to heat and illuminate objects; with remarks, that prove the different refrangibility of radiant heat. To which is added, an inquiry into the method of viewing the sun advantageously, with telescopes of large apertures and high magnifying powers. In Philosophical Transactions of the Royal Society of London 90 (XIII), 255-283. DOI: 10.1098/rstl.1800.0014.

Hertog, Maarten L. A. T. M.; Uysal, Ismail; McCarthy, Ultan; Verlinden, Bert M.; Nicolaï, Bart M. (2014): Shelf life modelling for first-expired-first-out warehouse management. In Philosophical transactions. Series A, Mathematical, physical, and engineering sciences 372 (20130306), pp. 1–15. DOI: 10.1098/rsta.2013.0306.

Hewett, Errol W. (2006): An overview of preharvest factors influencing postharvest quality of horticultural products. In International Journal of Postharvest Technology and Innovation 1 (1), pp. 4–15. DOI: 10.1504/IJPTI.2006.009178.

Hu, Rong; Zhang, Lixin; Yu, Zhiyuan; Zhai, Zhiqiang; Zhang, Ruoyu (2019): Optimization of soluble solids content prediction models in 'Hami' melons by means of Vis-NIR spectroscopy and chemometric tools. In Infrared Physics & Technology 102, p. 102999. DOI: 10.1016/j.infrared.2019.102999.

Huang, Haibo; Yu, Haiyan; Xu, Huirong; Ying, Yibin (2008): Near infrared spectroscopy for on/in-line monitoring of quality in foods and beverages: A review. In Journal of Food Engineering 87 (3), pp. 303–313. DOI: 10.1016/j.jfoodeng.2007.12.022.

Huang, Lingxia; Meng, Liuwei; Zhu, Nan; Wu, Di (2017): A primary study on forecasting the days before decay of peach fruit using near-infrared spectroscopy and electronic nose techniques. In Postharvest Biology and Technology 133, pp. 104–112. DOI: 10.1016/j.postharvbio.2017.07.014.

Hwang, Jay (2018): Sunforest H-100F. Written message to Simon Goisser. Freising, 10/15/2018. E-Mail.

Illert, Sonja (2019): Mehr Gemüse unter Glas angebaut. Edited by Agrarmarkt Informations-Gesellschaft mbH (AMI). Available online at https://www.ami-informiert.de/ami-maerkte/maerkte/ami-gartenbau/ami-meldungen-gartenbau, checked on 11/5/2020.

Inarejos-García, A. M.; Gómez-Alonso, S.; Fregapane, G.; Salvador, M. D. (2013): Evaluation of minor components, sensory characteristics and quality of virgin olive oil by near infrared (NIR) spectroscopy. In Food Research International 50 (1), pp. 250–258. DOI: 10.1016/j.foodres.2012.10.029.

Jantra, Chaiya; Slaughter, David C.; Liang, Pei-Shih; Pathaveerat, Siwalak (2017): Nondestructive determination of dry matter and soluble solids content in dehydrator onions and garlic using a handheld visible and near infrared instrument. In Postharvest Biology and Technology 133, pp. 98–103. DOI: 10.1016/j.postharvbio.2017.07.007.

Javanmardi, Jamal; Kubota, Chieri (2006): Variation of lycopene, antioxidant activity, total soluble solids and weight loss of tomato during postharvest storage. In Postharvest Biology and Technology 41 (2), pp. 151–155. DOI: 10.1016/j.postharvbio.2006.03.008.

Joachim, Verena; Spieth, Patrick; Heidenreich, Sven (2018): Active innovation resistance: An empirical study on functional and psychological barriers to innovation adoption in different contexts. In Industrial Marketing Management 71, pp. 95–107. DOI: 10.1016/J.INDMARMAN.2017.12.011.

Kaur, Harpreet; Künnemeyer, Rainer; McGlone, Andrew (2017): Comparison of hand-held near infrared spectrophotometers for fruit dry matter assessment. In Journal of Near Infrared Spectroscopy 25 (4), pp. 267–277. DOI: 10.1177/0967033517725530.

Kawano, Sumio; Fujiwara, Takayuki; Iwamoto, Mutsuo (1993): Nondestructive Determination of Sugar Content in Satsuma Mandarin using Near Infrared (NIR) Transmittance. In Journal of the Japanese Society for Horticultural Science 62 (2), pp. 465–470. DOI: 10.2503/jjshs.62.465.

Kawano, Sumio; Watanabe, Hisayoshi; Iwamoto, Mutsuo (1992): Determination of Sugar Content in Intact Peaches by Near Infrared Spectroscopy with Fiber Optics in Interactance Mode. In Journal of the Japanese Society for Horticultural Science 61 (2), pp. 445–451. DOI: 10.2503/jjshs.61.445.

Khatiwada, Bed P.; Subedi, Phul P.; Hayes, Clinton; Carlos, L. CunhaC.; Walsh, Kerry B. (2016): Assessment of internal flesh browning in intact apple using visible-short wave near infrared spectroscopy. In Postharvest Biology and Technology 120, pp. 103–111. DOI: 10.1016/j.postharvbio.2016.06.001.

Khodabakhshian, Rasool; Emadi, Bagher; Khojastehpour, Mehdi; Golzarian, Mahmood Reza; Sazgarnia, Ameneh (2017): Non-destructive evaluation of maturity and quality parameters of pomegranate fruit by visible/near infrared spectroscopy. In International Journal of Food Properties 20 (1), pp. 41–52. DOI: 10.1080/10942912.2015.1126725.

KOB (2020): Melon: Entwicklung einer Methode zur Bestimmung des optimalen Erntezeitpunkts anhand optischer Eigenschaften. Kompetenzzentrum Obstbau-Bavendorf. Available online at https://www.kob-bavendorf.de/projekte/laufende-projekte/melon-entwicklung-einer-methode-zur-bestimmung-des-optimalen-erntezeitpunkts-anhand-optischer-eigenschaften, checked on 12/16/2020.

Kong, F.; Singh, R. P. (2016): Chemical Deterioration and Physical Instability of Foods and Beverages. In : The Stability and Shelf Life of Food. Second Edition: Elsevier, pp. 43–76.

Ku, H. S.; Yang, S. F.; Pratt, H. K. (1970): Ethylene production and peroxidase activity during tomato fruit ripening. In Plant and Cell Physiology 11 (2), pp. 241–246. DOI: 10.1093/oxfordjournals.pcp.a074505.

Kubota (2020): Technische Parameter (deutsche Übersetzung via Google-Translator). Available online at https://fruitselector.com, checked on 10/20/2020.

Kumar, Satish; McGlone, Andrew; Whitworth, Claire; Volz, Richard (2015): Postharvest performance of apple phenotypes predicted by near-infrared (NIR) spectral analysis. In Postharvest Biology and Technology 100, pp. 16–22. DOI: 10.1016/j.postharvbio.2014.09.021.

Kusumiyati; Akinaga, Takayoshi; Tanaka, Munehiro; Kawasaki, Sheishi (2008): On-tree and after-harvesting evaluation of firmness, color and lycopene content of tomato fruit using portable NIR spectroscopy. In Journal of Food, Agriculture & Environment 6 (2), pp. 327–332.

Kusumiyati; Hamdani, J. S.; Sutari, W.; Mubarok, S.; Kurniasari, I. (2020): Non-destructive detection of two cucumber cultivars fruit quality using NIR Spectroscopy. In IOP Conference Series: Earth and Environmental Science 583, p. 12002. DOI: 10.1088/1755-1315/583/1/012002.

Kusumiyati; Mubarok, S.; Sutari, W.; Farida; Hamdani, J. S.; Hadiwijaya, Y.; Putri, I. E. (2019): Non-Destructive Method for Predicting Sapodilla Fruit Quality Using Near Infrared Spectroscopy. In IOP Conference Series: Earth and Environmental Science 334, p. 12045. DOI: 10.1088/1755-1315/334/1/012045.

Kusumiyati; Mubarok, Syariful; Hamdani, Jajang Sauman; Farida; Sutari, Wawan; Hadiwijaya, Yuda et al. (2018a): Evaluation of sapodilla fruit quality using near-infrared spectroscopy. In Journal of Food, Agriculture & Environment 16 (1), pp. 49–53. DOI: 10.31227/osf.io/mghf2.

Kusumiyati; Mubarok, Syariful; Hamdani, Jajang Sauman; Farida; Sutari, Wawan; Hadiwijaya, Yuda; Putri, Ine Elisa (2017): Detection of Ridge Gourd (Luffa acutangula) Fruit Quality During Storage Using Near-Infrared Spectrometer. In Proceeding of ISAE International Seminar, Bandar Lampung, August 10-12, 2017, 225-238.

Kusumiyati; Sutari, Wawan; Sauman, Jajang; Mubarok, Syariful; Sitepu, Rika Bhernike; Oktavia, Ade Risti (2018b): Non-Destructive Measurement of Green Bitter Gourd Quality Component Using Near Infrared Spectroscopy (NIRS). In Science and Technology Indonesia 3 (2), pp. 59–65. DOI: 10.26554/sti.2018.3.2.59-65.

Lammertyn, J.; Peirs, A.; De Baerdemaeker, J.; Nicolaï, B. M. (2001): Reply to the Letter to the Editor. NIR light penetration into fruit: a critical appraisal of measurement methodologies. In Postharvest Biology and Technology 22 (3), 194-195. DOI: 10.1016/S0925-5214(01)00104-1.

Lammertyn, Jeroen; Peirs, Ann; Baerdemaeker, Josse de; Nicolaï, Bart (2000): Light penetration properties of NIR radiation in fruit with respect to non-destructive quality assessment. In Postharvest Biology and Technology 18 (2), pp. 121–132. DOI: 10.1016/S0925-5214(99)00071-X.

Langer, Robert (2019): Foodscanning - Der Blickwinkel eines Technologieanbieters. Edited by Senorics.

Läpple, Doris; Renwick, Alan; Thorne, Fiona (2015): Measuring and understanding the drivers of agricultural innovation: Evidence from Ireland. In Food Policy 51, pp. 1–8. DOI: 10.1016/j.foodpol.2014.11.003.

Lee, Seoho; Noh, Tae Gyoon; Choi, Jun Hoe; Han, Jeongsu; Ha, Joo Young; Lee, Ji Young; Park, Yongjong (2017): NIR spectroscopic sensing for point-of-need freshness assessment of meat, fish, vegetables and fruits. In Moon S. Kim, Kuanglin Chao, Bryan A. Chin, Byoung-Kwan Cho (Eds.): Sensing for Agriculture and Food Quality and Safety IX, vol. 10217. SPIE Commercial + Scientific Sensing and Imaging. Anaheim, California, United States, Sunday 9 April 2017: SPIE (SPIE Proceedings), p. 1021708.

Li, Ming; Lv, Wenbo; Zhao, Rui; Guo, Huixin; Liu, Jing; Han, Donghai (2017a): Non-destructive assessment of quality parameters in 'Friar' plums during low temperature storage using visible/near infrared spectroscopy. In Food Control 73, pp. 1334–1341. DOI: 10.1016/j.foodcont.2016.10.054.

Li, Mo; Qian, Zhiqing; Shi, Bingwen; Medlicott, Jane; East, Andrew (2018): Evaluating the performance of a consumer scale SCiO™ molecular sensor to predict quality of horticultural products. In Postharvest Biology and Technology 145, pp. 183–192. DOI: 10.1016/j.postharvbio.2018.07.009.

Li, Tianhua; Zhong, Chongzhe; Lou, Wei; Wei, Min; Hou, Jialin (2017b): Optimization of Characteristic Wavelengths in Prediction of Lycopene in Tomatoes Using Near-Infrared Spectroscopy. In Journal of Food Process Engineering 40 (1), e12266. DOI: 10.1111/jfpe.12266.

Liu, Ran; Qi, Shuye; Lu, Jie; Han, Donghai (2015): Measurement of Soluble Solids Content of Three Fruit Species Using Universal near Infrared Spectroscopy Models. In Journal of Near Infrared Spectroscopy 23 (5), pp. 301–309. DOI: 10.1255/jnirs.1156.

Lu, Jie; Qi, Shuye; Liu, Ran; Zhou, Enyang; Li, Wu; Song, Shuhui; Han, Donghai (2015): Nondestructive determination of soluble solids and firmness in mix-cultivar melon using near-infrared CCD spectroscopy. In Journal of Innovative Optical Health Sciences 08 (06), p. 1550032. DOI: 10.1142/S1793545815500327.

Lu, Xiaonan; Wang, Jun; Al-Qadiri, Hamzah M.; Ross, Carolyn F.; Powers, Joseph R.; Tang, Juming; Rasco, Barbara A. (2011): Determination of total phenolic content and antioxidant capacity of onion (Allium cepa) and shallot (Allium oschaninii) using infrared spectroscopy. In Food Chemistry 129 (2), pp. 637–644. DOI: 10.1016/j.foodchem.2011.04.105.

Ma, Yizhou B.; Babu, Karthik S.; Amamcharla, Jayendra K. (2019): Prediction of total protein and intact casein in cheddar cheese using a low-cost handheld short-wave near-infrared spectrometer. In LWT - Food Science and Technology 109, pp. 319–326. DOI: 10.1016/j.lwt.2019.04.039.

Magwaza, Lembe Samukelo; Opara, Umezuruike Linus; Cronje, Paul J.R.; Landahl, Sandra; Nieuwoudt, Hélène H.; Mouazen, Abdul M. et al. (2014): Assessment of rind quality of 'Nules Clementine' mandarin fruit during postharvest storage: 2. Robust Vis/NIRS PLS models for prediction of physico-chemical attributes. In Scientia Horticulturae 165, pp. 421–432. DOI: 10.1016/j.scienta.2013.09.050.

Majidi, H.; Minaei, S.; Almassi, M.; Mostofi, Y. (2014): Tomato quality in controlled atmosphere storage, modified atmosphere packaging and cold storage. In Journal of Food Science and Technology 51 (9), pp. 2155–2161. DOI: 10.1007/s13197-012-0721-0.

Malegori, Cristina; Nascimento Marques, Emanuel José; Freitas, Sergio Tonetto de; Pimentel, Maria Fernanda; Pasquini, Celio; Casiraghi, Ernestina (2017): Comparing the analytical performances of Micro-NIR and FT-NIR spectrometers in the evaluation of acerola fruit quality, using PLS and SVM regression algorithms. In Talanta 165, pp. 112–116. DOI: 10.1016/j.talanta.2016.12.035.

Mani, Zied; Chouk, Inès (2017): Drivers of consumers' resistance to smart products. In Journal of Marketing Management 33, pp. 76–97. DOI: 10.1080/0267257X.2016.1245212.

Marques, Emanuel José Nascimento; Freitas, Sérgio Tonetto de (2020): Performance of new low-cost handheld NIR spectrometers for nondestructive analysis of umbu (Spondias tuberosa Arruda) quality. In Food Chemistry 323, p. 126820. DOI: 10.1016/j.foodchem.2020.126820.

Marques, Emanuel José Nascimento; Freitas, Sérgio Tonetto de; Pimentel, Maria Fernanda; Pasquini, Celio (2016): Rapid and non-destructive determination of quality parameters in the 'Tommy Atkins' mango using a novel handheld near infrared spectrometer. In Food Chemistry 197, pp. 1207–1214. DOI: 10.1016/j.foodchem.2015.11.080.

Martínez-Romero, Domingo; Bailén, Gloria; Serrano, María; Guillén, Fabián; Valverde, Juan Miguel; Zapata, Pedro et al. (2007): Tools to Maintain Postharvest Fruit and Vegetable Quality through the Inhibition of Ethylene Action: A Review. In Critical Reviews in Food Science and Nutrition 47 (6), pp. 543–560. DOI: 10.1080/10408390600846390.

Mazivila, Sarmento J.; Páscoa, Ricardo N. M. J.; Castro, Rafael C.; Ribeiro, David S. M.; Santos, João L. M. (2020): Detection of melamine and sucrose as adulterants in milk powder using near-infrared spectroscopy with DD-SIMCA as one-class classifier and MCR-ALS as a means to provide pure profiles of milk and of both adulterants with forensic evidence: A short communication. In Talanta 216, p. 120937. DOI: 10.1016/j.talanta.2020.120937.

McClure, W. F.; Norris, K. H.; Weeks, W. W. (1977): Rapid Spectrophotometric Analysis of the Chemical Composition of Tobacco. Part 1: Total Reducing Sugars. In Beiträge zur Tabakforschung 9 (1), pp. 13–18. DOI: 10.2478/cttr-2013-0421.

McClure, W. Fred (1994): Near-Infrared Spectroscopy. The Giant is Running Strong. In Analytical Chemistry 66 (1), pp. 43–53. DOI: 10.1021/ac00073a002.

McGlone, V.Andrew; Jordan, Robert B.; Martinsen, Paul J. (2002): Vis/NIR estimation at harvest of pre- and post-storage quality indices for 'Royal Gala' apple. In Postharvest Biology and Technology 25 (2), pp. 135–144. DOI: 10.1016/S0925-5214(01)00180-6.

McIntyre, Samantha M.; Ma, Qianli; Burritt, David J.; Oey, Indrawati; Gordon, Keith C.; Fraser-Miller, Sara J. (2020): Vibrational spectroscopy and chemometrics for quantifying key bioactive components of various plum cultivars grown in New Zealand. In Journal of Raman Spectroscopy 51 (7), pp. 1138–1152. DOI: 10.1002/jrs.5867.

Mehinagic, Emira; Royer, Gaëlle; Bertrand, Dominique; Symoneaux, Ronan; Laurens, François; Jourjon, Frédérique (2003): Relationship between sensory analysis, penetrometry and visible–NIR spectroscopy of apples belonging to different cultivars. In Food Quality and Preference 14 (5-6), pp. 473–484. DOI: 10.1016/S0950-3293(03)00012-0.

Minas, Ioannis S.; Blanco-Cipollone, Fernando; Sterle, David (2021): Accurate non-destructive prediction of peach fruit internal quality and physiological maturity with a single scan using near infrared spectroscopy. In Food Chemistry 335, p. 127626. DOI: 10.1016/j.foodchem.2020.127626.

Mishra, Puneet; Marini, Federico; Brouwer, Bastiaan; Roger, Jean Michel; Biancolillo, Alessandra; Woltering, Ernst; Echtelt, Esther Hogeveen-van (2021a): Sequential fusion of information from two portable spectrometers for improved prediction of moisture and soluble solids content in pear fruit. In Talanta 223, p. 121733. DOI: 10.1016/j.talanta.2020.121733.

Mishra, Puneet; Woltering, Ernst; Brouwer, Bastiaan; Echtelt, Esther Hogeveen-van (2021b): Improving moisture and soluble solids content prediction in pear fruit using near-infrared spectroscopy with variable selection and model updating approach. In Postharvest Biology and Technology 171, p. 111348. DOI: 10.1016/j.postharvbio.2020.111348.

Mishra, Puneet; Woltering, Ernst; El Harchioui, Najim (2020): Improved prediction of 'Kent' mango firmness during ripening by near-infrared spectroscopy supported by interval partial least square regression. In Infrared Physics & Technology 110, p. 103459. DOI: 10.1016/j.infrared.2020.103459.

Mithun, B S; Mondal, Milton; Vishwakarma, Harsh; Shinde, Sujit; Kimbahune, Sanjay (2017): Detection of artificially ripened mango using spectrometric analysis. In Moon S. Kim, Kuanglin Chao, Bryan A. Chin, Byoung-Kwan Cho (Eds.): Sensing for Agriculture and Food Quality and Safety IX. SPIE Commercial + Scientific Sensing and Imaging. Anaheim, California, United States, Sunday 9 April 2017: SPIE (SPIE Proceedings), 102170A.

Mogollón, René; Contreras, Carolina; Da Silva Neta, Magnólia Lourenço; Marques, Emanuel José Nascimento; Zoffoli, Juan Pablo; Freitas, Sergio Tonetto de (2020): Non-destructive prediction and detection of internal physiological disorders in 'Keitt' mango using a hand-held Vis-NIR spectrometer. In Postharvest Biology and Technology 167, p. 111251. DOI: 10.1016/j.postharvbio.2020.111251.

Momin, Md Abdul; Valero Ubierna, Constantino; Donis-Gonzalez, Irwin R.; Bergman, Shira; Slaughter, David C. (2019): Evaluación metrológica de dos dispositivos comerciales para la estimación espectrofotométrica de la calidad interna en uvas. In Pablo Martín Ramos, Francisco Javier García Ramos (Eds.): Congreso Ibérico de Agroingeniería. X Congreso

Ibérico de Agroingeniería = X Congresso Ibérico de Agroengenharia. Huesca, Spain, 03.09.2019 - 06.09.2019. Zaragoza: Servicio de Publicaciones Universidad, pp. 1227–1237.

Næs, Tormod; Isaksson, Tomas; Fearn, Tom; Davies, Tony (2004): A user-friendly guide to Multivariate Calibration and Classification. Chichester, UK: NIR Publications.

Ncama, Khayelihle; Magwaza, Lembe S.; Poblete-Echeverría, Carlos A.; Nieuwoudt, Hélène H.; Tesfay, Samson Z.; Mditshwa, Asanda (2018): On-tree indexing of 'Hass' avocado fruit by non-destructive assessment of pulp dry matter and oil content. In Biosystems Engineering 174, pp. 41–49. DOI: 10.1016/j.biosystemseng.2018.06.011.

Nicolaï, Bart M.; Beullens, Katrien; Bobelyn, Els; Peirs, Ann; Saeys, Wouter; Theron, Karen I.; Lammertyn, Jeroen (2007): Nondestructive measurement of fruit and vegetable quality by means of NIR spectroscopy: A review. In Postharvest Biology and Technology 46 (2), pp. 99–118. DOI: 10.1016/j.postharvbio.2007.06.024.

Norris, K. H.; Barnes, R. F.; Moore, J. E.; Shenk, J. S. (1976): Predicting Forage Quality by Infrared Reflectance Spectroscopy. In Journal of Animal Science 43 (4), pp. 889–897. DOI: 10.2527/jas1976.434889x.

Norris, Karl H.; Hart, Joe R. (1965): Direct Spectrophotometric Determination of Moisture Content of Grain and Seeds. In Proceedings of the 1963 International Symposium on Humidity and Moisture, Principles and Methods of Measuring Moisture in Liquids and Solids 4, pp. 19–25.

OECD (2018): OECD Fruit and Vegetalbe scheme. Guidelines on objective tests to determine quality of fruit and vegetables, dry and dried produce. Available online at http://www.oecd.org/agriculture/fruit-vegetables/, checked on 9/16/2020.

Olarewaju, Olaoluwa Omoniyi; Bertling, Isa; Magwaza, Lembe Samukelo (2016): Non-destructive evaluation of avocado fruit maturity using near infrared spectroscopy and PLS regression models. In Scientia Horticulturae 199, pp. 229–236. DOI: 10.1016/j.scienta.2015.12.047.

Oliveira, Gabrieli Alves de; Bureau, Sylvie; Renard, Catherine Marie-Geneviève Claire; Pereira-Netto, Adaucto Bellarmino; Castilhos, Fernanda de (2014): Comparison of NIRS approach for prediction of internal quality traits in three fruit species. In Food Chemistry 143, pp. 223–230. DOI: 10.1016/j.foodchem.2013.07.122.

Oms-Oliu, Gemma; Hertog, M. L. A. T. M.; van de Poel, B.; Ampofo-Asiama, J.; Geeraerd, A. H.; Nicolaï, B. M. (2011): Metabolic characterization of tomato fruit during preharvest development, ripening, and postharvest shelf-life. In Postharvest Biology and Technology 62 (1), pp. 7–16. DOI: 10.1016/j.postharvbio.2011.04.010.

Oravec, Michal; Beganović, Anel; Gál, Lukáš; Čeppan, Michal; Huck, Christian W. (2019): Forensic classification of black inkjet prints using Fourier transform near-infrared spectroscopy and Linear Discriminant Analysis. In Forensic Science International 299, pp. 128–134. DOI: 10.1016/j.forsciint.2019.03.041.

Osborne, Brian G. (2006): Near-Infrared Spectroscopy in Food Analysis. In Robert A. Meyers (Ed.): Encyclopedia of Analytical Chemistry. Chichester, UK: John Wiley & Sons, Ltd.

Parpinello, Giuseppina Paola; Nunziatini, Giulia; Rombolà, Adamo Domenico; Gottardi, Fernando; Versari, Andrea (2013): Relationship between sensory and NIR spectroscopy in consumer preference of table grape (cv Italia). In Postharvest Biology and Technology 83, pp. 47–53. DOI: 10.1016/j.postharvbio.2013.03.013.

Pasquini, Celio (2003): Near Infrared Spectroscopy: Fundamentals, Practical Aspects and Analytical Applications. In Journal of the Brazilian Chemical Society 14 (2), pp. 198–218. DOI: 10.1590/S0103-50532003000200006.

Pasquini, Celio (2018): Near infrared spectroscopy: A mature analytical technique with new perspectives - A review. In Analytica chimica acta 1026, pp. 8–36. DOI: 10.1016/j.aca.2018.04.004.

Pérez-Marín, Dolores; Calero, Luis; Fearn, Tom; Torres, Irina; Garrido-Varo, Ana; Sánchez, María-Teresa (2019a): A system using in situ NIRS sensors for the detection of product failing to meet quality standards and the prediction of optimal postharvest shelf-life in the case of oranges kept in cold storage. In Postharvest Biology and Technology 147, pp. 48–53. DOI: 10.1016/j.postharvbio.2018.09.009.

Pérez-Marín, Dolores; Sánchez, María-Teresa; Paz, Patricia; Soriano, María-Auxiliadora; Guerrero, José-Emilio; Garrido-Varo, Ana (2009): Non-destructive determination of quality parameters in nectarines during on-tree ripening and postharvest storage. In Postharvest Biology and Technology 52 (2), pp. 180–188. DOI: 10.1016/j.postharvbio.2008.10.005.

Pérez-Marín, Dolores; Torres, Irina; Entrenas, José-Antonio; Vega, Miguel; Sánchez, María-Teresa (2019b): Pre-harvest screening on-vine of spinach quality and safety using NIRS technology. In Spectrochimica Acta Part A: Molecular and Biomolecular Spectroscopy 207, pp. 242–250. DOI: 10.1016/j.saa.2018.09.035.

Pierpaoli, Emanuele; Carli, Giacomo; Pignatti, Erika; Canavari, Maurizio (2013): Drivers of Precision Agriculture Technologies Adoption: A Literature Review. In Procedia Technology 8, pp. 61–69. DOI: 10.1016/j.protcy.2013.11.010.

Pissard, Audrey; Fernández Pierna, Juan A.; Baeten, Vincent; Sinnaeve, Georges; Lognay, Georges; Mouteau, Anne et al. (2013): Non-destructive measurement of vitamin C, total polyphenol and sugar content in apples using near-infrared spectroscopy. In Journal of the Science of Food and Agriculture 93 (2), pp. 238–244. DOI: 10.1002/jsfa.5779.

Pissard, Audrey; Marques, Emanuel José Nascimento; Dardenne, Pierre; Lateur, Marc; Pasquini, Celio; Pimentel, Maria Fernanda et al. (2021): Evaluation of a handheld ultra-compact NIR spectrometer for rapid and non-destructive determination of apple fruit quality. In Postharvest Biology and Technology 172, p. 111375. DOI: 10.1016/j.postharvbio.2020.111375.

Polychromix (2007): Introducing the Polychromix PHAZIR Carpet Identification Tool. Available online at https://carpetrecovery.org/wp-content/uploads/2014/04/Phazir_CARE.pdf, checked on 12/18/2020.

Pöpping, Bert; Bourdichon, François (2018): Consumer food testing devices: threat or opportunity? In New Food 21 (1), pp. 30–33. Available online at newfoodmagazine.com.

Posom, Jetsada; Soonnamtiang, Navavit; Kotethum, Patcharapong; Konjun, Pakhpoom; Sirisomboon, Panmanas; Saengprachatanarug, Khwantri; Wongpichet, Seree (2020): Two Different Portables Visible-Near Infrared and Shortwave Infrared Region for On-Tree Measurement of Soluble Solid Content of Marian Plum Fruit. In Engineering Journal 24 (5), pp. 227–236. DOI: 10.4186/ej.2020.24.5.227.

Pu, Yuan-Yuan; Sun, Da-Wen; Riccioli, Cecilia; Buccheri, Marina; Grassi, Maurizio; Cattaneo, Tiziana M. P.; Gowen, Aoife (2018): Calibration Transfer from Micro NIR Spectrometer to Hyperspectral Imaging: a Case Study on Predicting Soluble Solids Content of Bananito Fruit (Musa acuminata). In Food Analytical Methods 11 (4), pp. 1021–1033. DOI: 10.1007/s12161-017-1055-3.

Qi, Shuye; Oshita, Seiichi; Makino, Yoshio; Han, Donghai (2017): Influence of Sampling Component on Determination of Soluble Solids Content of Fuji Apple Using Near-Infrared Spectroscopy. In Applied spectroscopy 71 (5), pp. 856–865. DOI: 10.1177/0003702816658671.

Rady, Ahmed; Fischer, Joel; Reeves, Stuart; Logan, Brian; Watson, Nicholas James (2020): The Effect of Light Intensity, Sensor Height, and Spectral Pre-Processing Methods when using NIR Spectroscopy to Identify Different Allergen-Containing Powdered Foods. In Sensors 20 (230), pp. 1–24. DOI: 10.3390/s20010230.

Radzevičius, Audrius; Viškelis, Jonas; Karklelienė, Rasa; Juškevičienė, Danguolė; Viškelis, Pranas (2016): Determination of tomato quality attributes using near infrared spectroscopy and reference analysis. In Zemdirbyste-Agriculture 103 (1), pp. 443–448. DOI: 10.13080/z-a.2016.103.012.

Rani, Monika; Marchesi, Claudio; Federici, Stefania; Rovelli, Gianluca; Alessandri, Ivano; Vassalini, Irene et al. (2019): Miniaturized Near-Infrared (MicroNIR) Spectrometer in Plastic Waste Sorting. In Materials 12 (17), pp. 1–13. DOI: 10.3390/ma12172740.

Rateni, Giovanni; Dario, Paolo; Cavallo, Filippo (2017): Smartphone-Based Food Diagnostic Technologies: A Review. In Sensors (Basel, Switzerland) 17 (6). DOI: 10.3390/s17061453.

Ribeiro, Lívia Paulia Dias; Da Silva, Ana Priscila Monteiro; Lima, Aliny Alencar de; Oliveira Silva, Ebenézer de; Rinnan, Åsmund; Pasquini, Celio (2016): Non-Destructive Determination of Quality Traits of Cashew Apples (Anacardium Occidentale, L.) Using a Portable near Infrared Spectrophotometer. In Journal of Near Infrared Spectroscopy 24 (1), pp. 77–82. DOI: 10.1255/jnirs.1197.

Risoluti, Roberta; Gregori, Adolfo; Schiavone, Sergio; Materazzi, Stefano (2018): "Click and Screen" Technology for the Detection of Explosives on Human Hands by a Portable MicroNIR-Chemometrics Platform. In Analytical Chemistry 90 (7), pp. 4288–4292. DOI: 10.1021/acs.analchem.7b03661.

Roussaki, Ioanna; Kosmides, Pavlos; Routis, George; Doolin, Kevin; Pevtschin, Veronique; Marguglio, Angelo (2019): A Multi-Actor Approach to promote the employment of IoT in Agriculture. In IEEE (Ed.): 2019 Global IoT Summit (GIoTS). 2019 Global IoT Summit (GIoTS). Aarhus, Denmark, 17.06.2019 - 21.06.2019: IEEE, pp. 1–6. DOI: 10.1109/GIOTS.2019.8766416

Roy, Soumya; Anantheswaran, Ramaswamy C.; Shenk, John S.; Westerhaus, Mark O.; Beelman, Robert B. (1993): Determination of Moisture Content of Mushrooms by Vis-NIR Spectroscopy. In Journal of the Science of Food and Agriculture 63 (3), pp. 355–360. DOI: 10.1002/jsfa.2740630314.

Sánchez, M. T.; Torres, I.; Gil, B.; Pérez-Marín, D.; Garrido-Varo, A.; La Haba, M. J. de (2018): In-situ determination of external quality parameters in intact summer squash using near-infrared reflectance spectroscopy. In Acta Horticulturae (1194), pp. 1259–1264. DOI: 10.17660/ActaHortic.2018.1194.178.

Sánchez, María-Teresa; Pérez-Marín, Dolores; Torres, Irina; Gil, Belén; Garrido-Varo, Ana; La Haba, María-José de (2017): Use of NIRS technology for on-vine measurement of nitrate content and other internal quality parameters in intact summer squash for baby food production. In Postharvest Biology and Technology 125, pp. 122–128. DOI: 10.1016/j.postharvbio.2016.11.011.

Sánchez, María-Teresa; Pintado, Carlos; La Haba, María-José de; Torres, Irina; García, Manuel; Pérez-Marín, Dolores (2020): In situ ripening stages monitoring of Lamuyo pepper using a new-generation near-infrared spectroscopy sensor. In Journal of the Science of Food and Agriculture 100 (5), pp. 1931–1939. DOI: 10.1002/jsfa.10205.

Sánchez, María-Teresa; Torres, Irina; La Haba, María-José de; Chamorro, Ana; Garrido-Varo, Ana; Pérez-Marín, Dolores (2019): Rapid, simultaneous, and in situ authentication and quality assessment of intact bell peppers using near-infrared spectroscopy technology. In Journal of the Science of Food and Agriculture 99 (4), pp. 1613–1622. DOI: 10.1002/jsfa.9342.

Sandorfy, C.; Buchet, R.; Lachenal, G. (2007): Principles of Molecular Vibrations for Near-Infrared Spectroscopy. In Y. Ozaki, W. F. McClure, Alfred A. Christy (Eds.): Near-infrared spectroscopy in food science and technology. Hoboken N.J.: Wiley-Interscience, pp. 11–46.

Santos, Carla S. P.; Cruz, Rebeca; Gonçalves, Diogo B.; Queirós, Rafael; Bloore, Mark; Kovács, Zoltán et al. (2020): Non-destructive measurement of the internal quality of citrus fruits using a portable NIR device. In Journal of AOAC INTERNATIONAL (qsaa115). DOI: 10.1093/jaoacint/qsaa115.

Saranwong, Innapa; Sornsrivichai, Jinda; Kawano, Sumio (2003): On-Tree Evaluation of Harvesting Quality of Mango Fruit Using a Hand-Held NIR Instrument. In Journal of Near Infrared Spectroscopy 11 (4), pp. 283–293. DOI: 10.1255/jnirs.374.

Saranwong, Sirinnapa; Kawano, Sumio (2007): Fruits and Vegetables. In Y. Ozaki, W. F. McClure, Alfred A. Christy (Eds.): Near-infrared spectroscopy in food science and technology. Hoboken N.J.: Wiley-Interscience, pp. 219–245.

Saranwong, Sirinnapa; Thanapase, Warunee; Suttiwijitpukdee, Nattaporn; Rittiron, Ronnarit; Kasemsumran, Sumaporn; Kawano, Sumio (2010): Applying near Infrared Spectroscopy to the Detection of Fruit Fly Eggs and Larvae in Intact Fruit. In Journal of Near Infrared Spectroscopy 18 (4), pp. 271–280. DOI: 10.1255/jnirs.886.

Sarkar, Shagor; Basak, Jayanta Kumar; Moon, Byeong Eun; Kim, Hyeon Tae (2020): A Comparative Study of PLSR and SVM-R with Various Preprocessing Techniques for the Quantitative Determination of Soluble Solids Content of Hardy Kiwi Fruit by a Portable Vis/NIR Spectrometer. In Foods 9 (8). DOI: 10.3390/foods9081078.

Scalisi, Alessio; O'Connell, Mark Glenn (2020): Application of visible/NIR spectroscopy for the estimation of soluble solids, dry matter and flesh firmness in stone fruits. In Journal of the Science of Food and Agriculture. DOI: 10.1002/jsfa.10832.

Schmutzler, Matthias; Huck, Christian W. (2016): Simultaneous detection of total antioxidant capacity and total soluble solids content by Fourier transform near-infrared (FT-NIR) spectroscopy: A quick and sensitive method for on-site analyses of apples. In Food Control 66, pp. 27–37. DOI: 10.1016/j.foodcont.2016.01.026.

Schulz, H.; Drews, H.-H.; Quilitzsch, R.; Krüger, H. (1998): Application of near Infrared Spectroscopy for the Quantification of Quality Parameters in Selected Vegetables and Essential Oil Plants. In Journal of Near Infrared Spectroscopy 6 (A), A125-A130. DOI: 10.1255/jnirs.179.

Senorics (2020): How it works. Our technology. Available online at https://senorics.com/our-technology/, updated on 10/30/2020.

Serra, Sara; Goke, Alex; Diako, Charles; Vixie, Beata; Ross, Carolyn; Musacchi, Stefano (2019): Consumer perception of d'Anjou pear classified by dry matter at harvest using near-infrared spectroscopy. In International Journal of Food Science & Technology 54 (6), pp. 2256–2265. DOI: 10.1111/ijfs.14140.

Shao, Yongni; He, Yong; Gómez, Antihus H.; Pereir, Annia G.; Qiu, Zhengjun; Zhang, Yun (2007): Visible/near infrared spectrometric technique for nondestructive assessment of tomato 'Heatwave' (Lycopersicum esculentum) quality characteristics. In Journal of Food Engineering 81 (4), pp. 672–678. DOI: 10.1016/j.jfoodeng.2006.12.026.

Shen, Fei; Zhang, Bin; Cao, Chongjiang; Jiang, Xuesong (2018): On-line discrimination of storage shelf-life and prediction of post-harvest quality for strawberry fruit by visible and near infrared spectroscopy. In Journal of Food Process Engineering 41 (e12866), 1-8. DOI: 10.1111/jfpe.12866.

Sheng, Ren; Cheng, Wu; Li, Huanhuan; Ali, Shujat; Agyekum, Akwasi Akomeah; Chen, Quansheng (2019): Model development for soluble solids and lycopene contents of cherry tomato at different temperatures using near-infrared spectroscopy. In Postharvest Biology and Technology 156, p. 110952. DOI: 10.1016/j.postharvbio.2019.110952.

Silva, Carolina Santos; Borba, Flávia de Souza Lins; Pimentel, Maria Fernanda; Pontes, Marcio José Coelho; Honorato, Ricardo Saldanha; Pasquini, Celio (2013): Classification of blue pen ink using infrared spectroscopy and linear discriminant analysis. In Microchemical Journal 109, pp. 122–127. DOI: 10.1016/j.microc.2012.03.025.

Sinelli, Nicoletta; Spinardi, Anna; Di Egidio, Valentina; Mignani, Ilaria; Casiraghi, Ernestina (2008): Evaluation of quality and nutraceutical content of blueberries (Vaccinium corymbosum L.) by near and mid-infrared spectroscopy. In Postharvest Biology and Technology 50 (1), pp. 31–36. DOI: 10.1016/j.postharvbio.2008.03.013.

Singh, Hardit; Sridhar, Aneesh; Saini, Simarjeet S. (2020): Ultra-Low-Cost Self-Referencing Multispectral Detector for Non-Destructive Measurement of Fruit Quality. In Food Analytical Methods 13 (10), pp. 1879–1893. DOI: 10.1007/s12161-020-01810-7.

Sirisomboon, Panmanas; Tanaka, Munehiro; Kojima, Takayuki; Williams, Phil (2012): Nondestructive estimation of maturity and textural properties on tomato 'Momotaro' by near infrared spectroscopy. In Journal of Food Engineering 112 (3), pp. 218–226. DOI: 10.1016/j.jfoodeng.2012.04.007.

Soderlund, Russell; Williams, Richard; Mulligan, Cathy (2008): Effective adoption of agri-food assurance systems. In British Food Journal 110 (8), pp. 745–761. DOI: 10.1108/00070700810893296.

Spectral Engines Oy (2018): Spectral Engines Webinar: Company presentation. Evolution and opportunities.

Spectral Engines Oy (2020): Food Scanner. The world's smartest, fastest and easiest way to measure food content. Available online at https://www.spectralengines.com/products/nirone-scanner/foodscanner, checked on 9/28/2020.

Statista (2020): Gemüsekonsum in Deutschland. Available online at https://de.statista.com/statistik/studie/id/42820/dokument/gemuesekonsum-in-deutschland/.

Su, Zheng; Lu, Zelong; Yang, Mingyang; Wu, Zhaoxin; Xu, Ruoyao (2020): Design and Experiment of a Portable Near-infrared Spectrum Detection System for Fruits and Vegetables Quality-inspection. In E3S Web of Conferences 189, p. 2009. DOI: 10.1051/e3sconf/202018902009.

Subedi, Phul P.; Walsh, Kerry B. (2020): Assessment of avocado fruit dry matter content using portable near infrared spectroscopy: Method and instrumentation optimisation. In Postharvest Biology and Technology 161, p. 111078. DOI: 10.1016/j.postharvbio.2019.111078.

Sukwanit, S.; Teerachaichayut, S. (2013): Nondestructive Prediction of Internal Browning in Pineapple Using Transmittance Short Wavelength Near Infrared Spectroscopy. In Acta Horticulturae (989), pp. 395–399. DOI: 10.17660/ActaHortic.2013.989.54.

Sun, Lan; Hsiung, Chang; Pederson, Christopher G.; Zou, Peng; Smith, Valton; Gunten, Marc von; O'Brien, Nada A. (2016): Pharmaceutical Raw Material Identification Using Miniature Near-Infrared (MicroNIR) Spectroscopy and Supervised Pattern Recognition Using Support Vector Machine. In Applied spectroscopy 70 (5), pp. 816–825. DOI: 10.1177/0003702816638281.

Sun, Xudong; Subedi, Phul; Walker, Rachel; Walsh, Kerry B. (2020a): NIRS prediction of dry matter content of single olive fruit with consideration of variable sorting for normalisation pre-treatment. In Postharvest Biology and Technology 163, p. 111140. DOI: 10.1016/j.postharvbio.2020.111140.

Sun, Xudong; Subedi, Phul; Walsh, Kerry B. (2020b): Achieving robustness to temperature change of a NIRS-PLSR model for intact mango fruit dry matter content. In Postharvest Biology and Technology 162, p. 111117. DOI: 10.1016/j.postharvbio.2019.111117.

Sunforest (2020): H-100 Series. Available online at http://sunforest.kr/category_main.php?sm_idx=167, checked on 9/28/2020.

Szuvandzsiev, Péter; Helyes, Lajos; Lugasi, Andrea; Szántó, Csongor; Baranowski, Piotr; Pék, Zoltán (2014): Estimation of antioxidant components of tomato using VIS-NIR reflectance data by handheld portable spectrometer. In International Agrophysics 28 (4). DOI: 10.2478/intag-2014-0042.

Tamburini, Elena; Costa, Stefania; Rugiero, Irene; Pedrini, Paola; Marchetti, Maria Gabriella (2017): Quantification of Lycopene, β-Carotene, and Total Soluble Solids in Intact Red-Flesh Watermelon (Citrullus lanatus) Using On-Line Near-Infrared Spectroscopy. In Sensors (Basel, Switzerland) 17 (4), pp. 1–12. DOI: 10.3390/s17040746.

Teh, Soon Li; Coggins, Jamie L.; Kostick, Sarah A.; Evans, Kate M. (2020): Location, year, and tree age impact NIR-based postharvest prediction of dry matter concentration for 58 apple accessions. In Postharvest Biology and Technology 166, p. 111125. DOI: 10.1016/j.postharvbio.2020.111125.

Tellspec Inc. (2020a): Empowering a Healthier World. with Real-Time Analysis Using Portable Low-Cost Sensors. Available online at https://tellspec.com/, checked on 9/28/2020.

Tellspec Inc. (2020b): Order - Tellspec Enterprise Sensor. Available online at https://tellspec.com/order/, checked on 12/18/2020.

Thermo Fisher Scientific (2020): microPHAZIR™ RX Analysator. Available online at https://www.thermofisher.com/order/catalog/product/MICROPHAZIRRX#/MICROPHAZIRRX, checked on 12/18/2020.

Toivonen, Peter M. A.; Batista, Adrian; Lannard, Brenda (2017): Development of a predictive model for 'Lapins' sweet cherry dry matter content using a visible/near infrared spectrometer and its potential application to other cultivars. In Canadian Journal of Plant Science 97, pp. 1030–1035. DOI: 10.1139/CJPS-2017-0013.

Toledo-Martín, Eva María; García-García, María Carmen; Font, Rafael; Moreno-Rojas, José Manuel; Gómez, Pedro; Salinas-Navarro, María; Del Río-Celestino, Mercedes (2016): Application of visible/near-infrared reflectance spectroscopy for predicting internal and external quality in pepper. In Journal of the Science of Food and Agriculture 96 (9), pp. 3114–3125. DOI: 10.1002/jsfa.7488.

Toor, Ramandeep K.; Savage, Geoffrey P. (2006): Changes in major antioxidant components of tomatoes during post-harvest storage. In Food Chemistry 99 (4), pp. 724–727. DOI: 10.1016/j.foodchem.2005.08.049.

Torres, Irina; Pérez-Marín, Dolores; La Haba, María-José de; Sánchez, María-Teresa (2017): Developing universal models for the prediction of physical quality in citrus fruits analysed on-tree using portable NIRS sensors. In Biosystems Engineering 153, pp. 140–148. DOI: 10.1016/j.biosystemseng.2016.11.007.

Torres, Irina; Sánchez, María-Teresa; Entrenas, José-Antonio; Garrido-Varo, Ana; Pérez-Marín, Dolores (2019a): Monitoring quality and safety assessment of summer squashes along the food supply chain using near infrared sensors. In Postharvest Biology and Technology 154, pp. 21–30. DOI: 10.1016/j.postharvbio.2019.04.015.

Torres, Irina; Sánchez, María-Teresa; Garrido-Varo, Ana; Pérez-Marín, Dolores C. (2020a): New generation NIRS sensors for quality and safety assurance in summer squashes along the food supply chain. In Moon S. Kim (Ed.): Sensing for Agriculture and Food Quality and Safety XII. 27 April-8 May 2020, Online Only, United States. Sensing for Agriculture and Food Quality and Safety XII. Online Only, United States, 4/27/2020 - 5/1/2020. Bellingham, Wash: SPIE (Proceedings of SPIE. 5200-, volume 11421), p. 16. Available online at https://www.spiedigitallibrary.org/conference-proceedings-of-spie/11421/2559104/New-generation-NIRS-sensors-for-quality-and-safety-assurance-along/10.1117/12.2559104.full.

Torres, Irina; Sánchez, María-Teresa; La Haba, María-José de; Pérez-Marín, Dolores (2019b): LOCAL regression applied to a citrus multispecies library to assess chemical quality parameters using near infrared spectroscopy. In Spectrochimica acta. Part A, Molecular and biomolecular spectroscopy 217, pp. 206–214. DOI: 10.1016/j.saa.2019.03.090.

Torres, Irina; Sánchez, María-Teresa; Pérez-Marín, Dolores (2020b): Integrated soluble solid and nitrate content assessment of spinach plants using portable NIRS sensors along the supply chain. In Postharvest Biology and Technology 168, p. 111273. DOI: 10.1016/j.postharvbio.2020.111273.

Torres, Irina; Sánchez, María-Teresa; Vega-Castellote, Miguel; Luqui-Muñoz, Natividad; Pérez-Marín, Dolores (2021): Routine NIRS analysis methodology to predict quality and safety indexes in spinach plants during their growing season in the field. In Spectrochimica Acta Part A: Molecular and Biomolecular Spectroscopy 246, p. 118972. DOI: 10.1016/j.saa.2020.118972.

trinamix GmbH (2020): Welcome to trinamiX. Available online at https://trinamixsensing.com/, checked on 12/17/2020.

Tsuchikawa, Satoru (2007): Sampling Techniques. In Y. Ozaki, W. F. McClure, Alfred A. Christy (Eds.): Near-infrared spectroscopy in food science and technology. Hoboken N.J.: Wiley-Interscience, pp. 133–143.

UNECE (2019): Fresh Fruit and Vegetables - Standards. Available online at http://www.unece.org/trade/agr/standard/fresh/ffv-standardse.html, checked on 9/16/2020.

Uwadaira, Yasuhiro; Sekiyama, Yasuyo; Ikehata, Akifumi (2018): An examination of the principle of non-destructive flesh firmness measurement of peach fruit by using VIS-NIR spectroscopy. In Heliyon 4 (2), e00531. DOI: 10.1016/j.heliyon.2018.e00531.

Valderrama, Patrícia; Braga, Jez Willian B.; Poppi, Ronei Jesus (2007): Variable selection, outlier detection, and figures of merit estimation in a partial least-squares regression multivariate calibration model. A case study for the determination of quality parameters in the alcohol industry by near-infrared spectroscopy. In Journal of Agricultural and Food Chemistry 55 (21), pp. 8331–8338. DOI: 10.1021/jf071538s.

VIAVI Solutions Inc. (2020): MicroNIR Spectrometers. Available online at https://www.viavisolutions.com/en-us/osp/products/micronir-spectrometers, checked on 12/18/2020.

Walsh, K. B. (2006): Setting and Meeting Objective Standards for Eating Quality in Fresh Fruit. In Acta Horticulturae (712), pp. 191–200. DOI: 10.17660/ActaHortic.2006.712.19.

Wang, Jiahua; Wang, Jun; Chen, Zhuo; Han, Donghai (2017): Development of multi-cultivar models for predicting the soluble solid content and firmness of European pear (Pyrus communis L.) using portable vis–NIR spectroscopy. In Postharvest Biology and Technology 129, pp. 143–151. DOI: 10.1016/j.postharvbio.2017.03.012.

Wang, Tao; Chen, Jian; Fan, Yangyang; Qiu, Zhengjun; He, Yong (2018): SeeFruits: Design and evaluation of a cloud-based ultra-portable NIRS system for sweet cherry quality detection. In Computers and Electronics in Agriculture 152, pp. 302–313. DOI: 10.1016/j.compag.2018.07.017.

Wedding, B. B.; Wright, C.; Grauf, S.; White, R. D.; Tilse, B.; Gadek, P. (2013): Effects of seasonal variability on FT-NIR prediction of dry matter content for whole Hass avocado fruit. In Postharvest Biology and Technology 75, pp. 9–16. DOI: 10.1016/j.postharvbio.2012.04.016.

Wiedemair, Verena; Huck, Christian W. (2018): Evaluation of the performance of three hand-held near-infrared spectrometer through investigation of total antioxidant capacity in gluten-free grains. In Talanta 189, pp. 233–240. DOI: 10.1016/j.talanta.2018.06.056.

Włodarska, Katarzyna; Szulc, Julia; Khmelinskii, Igor; Sikorska, Ewa (2019): Non-destructive determination of strawberry fruit and juice quality parameters using ultraviolet, visible, and near-infrared spectroscopy. In Journal of the Science of Food and Agriculture 99 (13), pp. 5953–5961. DOI: 10.1002/jsfa.9870.

Wokadala, Obiro Cuthbert; Human, Christo; Willemse, Salomie; Emmambux, Naushad Mohammad (2020): Rapid non-destructive moisture content monitoring using a handheld portable Vis–NIR spectrophotometer during solar drying of mangoes (Mangifera indica L.). In Journal of Food Measurement and Characterization 14 (2), pp. 790–798. DOI: 10.1007/s11694-019-00327-w.

Xiaobo, Zou; Jiewen, Zhao; Povey, Malcolm J. W.; Holmes, Mel; Hanpin, Mao (2010): Variables selection methods in near-infrared spectroscopy. In Analytica chimica acta 667 (1-2), pp. 14–32. DOI: 10.1016/j.aca.2010.03.048.

Xing, Juan; van Linden, Veerle; Vanzeebroeck, Michael; Baerdemaeker, Josse de (2005): Bruise detection on Jonagold apples by visible and near-infrared spectroscopy. In Food Control 16 (4), pp. 357–361. DOI: 10.1016/j.foodcont.2004.03.016.

Yan, Hui; Xu, Yi-Chao; Siesler, Heinz W.; Han, Bang-Xing; Zhang, Guo-Zheng (2019): Hand-Held Near-Infrared Spectroscopy for Authentication of Fengdous and Quantitative Analysis of Mulberry Fruits. In Frontiers in plant science 10, p. 1548. DOI: 10.3389/fpls.2019.01548.

Yang, I-Chang; Tsai, Chao-Yin; Hsieh, Kuang-Wen; Yang, Ci-Wen; Ouyang, Fu; Martin Lo, Yangming; Chen, Suming (2013): Integration of SIMCA and near-infrared spectroscopy for rapid and precise identification of herbal medicines. In Journal of Food and Drug Analysis 21 (3), pp. 268–278. DOI: 10.1016/j.jfda.2013.07.008.

Yang, Junxing; Lou, Xiaoping; Yang, Hongqi; Yang, Huibing; Liu, Chaoying; Wu, Jingjing; Bin, Jun (2019): Improved calibration transfer between near-Infrared (NIR) spectrometers using canonical correlation analysis. In Analytical Letters 52 (14), pp. 2188–2202. DOI: 10.1080/00032719.2019.1604725.

You, Xinge; Mou, Yi; Yu, Shujian; Jiang, Xiubao; Xu, Duanquan; Zhou, Long (2016): Mixed-Norm Partial Least Squares. In Chemometrics and Intelligent Laboratory Systems 152, pp. 42–53. DOI: 10.1016/j.chemolab.2016.01.004.

Yu, Xiang; Liu, Qing; Wang, Yebao; Liu, Xiangyang; Liu, Xin (2016): Evaluation of MLSR and PLSR for estimating soil element contents using visible/near-infrared spectroscopy in apple orchards on the Jiaodong peninsula. In CATENA 137, pp. 340–349. DOI: 10.1016/j.catena.2015.09.024.

Yuan, Lei-Ming; Mao, Fei; Chen, Xiaojing; Li, Limin; Huang, Guangzao (2020): Non-invasive measurements of 'Yunhe' pears by vis-NIRS technology coupled with deviation fusion modeling approach. In Postharvest Biology and Technology 160, p. 111067. DOI: 10.1016/j.postharvbio.2019.111067.

ZDNet (2017): Changhong präsentiert mit dem H2 erstes Smartphone mit integriertem Molekularscanner. Available online at https://www.zdnet.de/88285948/changhong-praesentiert-mit-dem-h2-erstes-smartphone-mit-integriertem-molekularscanner/, updated on 11/19/2020.

Zhang, Baohua; Dai, Dejian; Huang, Jichao; Zhou, Jun; Gui, Qifa (2017): Influence of physical and biological variability and solution methods in fruit and vegetable quality nondestructive inspection by using imaging and near-infrared spectroscopy techniques: A review. In Critical Reviews in Food Science and Nutrition 58 (12), pp. 2099–2118. DOI: 10.1080/10408398.2017.1300789.

Zhang, Shujuan; Zhang, Haihong; Zhao, Yanru; Guo, Wei; Zhao, Huamin (2013): A simple identification model for subtle bruises on the fresh jujube based on NIR spectroscopy. In Mathematical and Computer Modelling 58 (3-4), pp. 545–550. DOI: 10.1016/j.mcm.2011.10.067.

Zhang, Y.; Nock, J. F.; Al Shoffe, Y.; Watkins, C. B. (2020): Non-destructive prediction of soluble solids and dry matter concentrations in apples using near-infrared spectroscopy. In Acta Hortic. (1275), pp. 341–348. DOI: 10.17660/ActaHortic.2020.1275.47.

Zhang, Yiyi; Nock, Jacqueline F.; Al Shoffe, Yosef; Watkins, Christopher B. (2019): Non-destructive prediction of soluble solids and dry matter contents in eight apple cultivars using near-infrared spectroscopy. In Postharvest Biology and Technology 151, pp. 111–118. DOI: 10.1016/j.postharvbio.2019.01.009.

Zhiming, Guo; Quansheng, Chen; Bin, Zhang; Qingyan, Wang; Qin, Ouyang; Jiewen, Zhao (2017): Design and experiment of handheld near-infrared spectrometer for determination of fruit and vegetable quality. In Transactions of the Chinese Society of Agricultural Engineering 33 (8), pp. 245–250. DOI: 10.11975/j.issn.1002-6819.2017.08.033.

Zhu, Xiangrong; Li, Shuifang; Shan, Yang; Zhang, Zhuoyong; Li, Gaoyang; Su, Donglin; Liu, Feng (2010): Detection of adulterants such as sweeteners materials in honey using near-infrared spectroscopy and chemometrics. In Journal of Food Engineering 101 (1), pp. 92–97. DOI: 10.1016/j.jfoodeng.2010.06.014.

Ziegeldorf, Jan Henrik; Morchon, Oscar Garcia; Wehrle, Klaus (2014): Privacy in the Internet of Things: Threats and Challenges. In Security and Communication Networks 7 (12), pp. 2728–2742. DOI: 10.1002/sec.795.

12 Appendix

Table 5: Search results of publications per year related to keywords including "**NIR**" and additional terms of fruit quality according to Google Scholar in December 2020

Term / Year	2006	2007	2008	2009	2010	2011	2012	2013	2014	2015	2016	2017	2018	2019
"NIR" + "fruit quality"	178	167	208	195	279	267	368	395	428	433	429	496	582	606
"NIR" + "fruit" + "sugar content"	192	170	176	185	233	214	285	317	299	318	296	358	362	432
"NIR" + "fruit" + "dry matter"	162	153	165	175	225	215	305	306	329	388	375	431	473	502
"NIR" + "fruit" + "firmness"	163	178	213	217	288	267	329	380	404	414	414	480	486	530
"NIR" + "fruit" + "acidity"	174	186	217	241	270	293	362	394	412	453	446	528	565	636

In der Reihe *Berliner ökophysiologische und phytomedizinische Schriften* sind bisher erschienen:

Band 01: Mohammad Mahir Uddin (2009)
Chemical ecology of mustard leaf beetle Phaedon cochleariae (F.).
ISBN 978-3-89959-848-3.

Band 02: Ilir Morina (2009)
Entwicklung von Verfahren zur Rekultivierung der Aschedeponie des Braunkohlekraftwerks in Prishtina (Kosovo).
ISBN 978-3-89959-872-8.

Band 03: Melanie Wiesner (2009)
Veränderungen gesundheitsrelevanter Inhaltsstoffe in *Parthenium hysterophorus* L. in Abhängigkeit von der Pflanzengröße und Klimafaktoren.
ISBN 978-3-89959-880-3.

Band 04: Fransika Rohr (2009)
Variabilität aliphatischer Glucosinolate in *Arabidopsis thaliana*-Ökotypen und deren Einfluss auf die Wirtspflanzeneignung von zwei folivoren Insektenarten.
ISBN 978-3-89959-884-9.

Band 05: Jutta Buchhop (2009)
Characterization of phylogenetically diverse CLRV-isolates by RFLP and research into identification of two isometric viruses.
ISBN 978-3-89959-929-9.

Band 06: Nora Koim (2010)
Urban sprawl, land cover change and forest fragmentation – Case study Pereira, Colombia.
ISBN 978-3-89959-955-8.

Band 07: Nadja Förster (2010)
Eignung unterschiedlicher salicylathaltiger *Salix*-Klone für die Arzneimittelindustrie.
ISBN 978-3-89959-964-0.

Band 08: Jana Gentkow (2010)
Cherry leaf roll virus (CLRV): Charakterisierung ausgewählter Virusisolate unter besonderer Berücksichtigung des viralen Hüllproteins.
ISBN 978-3-89959-976-3.

Band 09: Ahmad Fakhro (2010)
Interaction of Pepino mosaic virus (PepMV) and fungal root endophytes with tomato hosts (*Lycopersicum esculentum* Mill.).
ISBN 978-3-89959-995-4.

Band 10: Stefan Irrgang (2010)
Mikro- und makroskopische Untersuchungen an Veredelungsstellen von Straßenbäumen im Hinblick auf die Beeinflussung ihrer Bruchsicherheit.
ISBN 978-3-89959-998-5.

Band 11: Julia Jahnke (2010)
Guerilla Gardening anhand von Beispielen in New York, London und Berlin.
ISBN 978-3-86247-001-3.

Band 12: Astrid Karoline Günther (2010)
Analysen zur Intensität der Pflanzenschutzmittel-Anwendung und Aufklärung ihrer Einflussfaktoren in ausgewählten Ackerbaubetrieben.
ISBN 978-3-86247-005-1.

Band 13: Milena A. Dimova (2010)
Untersuchungen zur Epidemiologie von *Pythium aphanidermatum* in Abhängigkeit von den Umgebungsbedingungen bei der Gewächshausgurke (*Cucumis sativus* L.).
ISBN 978-3-86247-033-4.

Band 14: Claudia Patricia Pérez-Rodríguez (2010)
Physiologische Veränderungen in Früchten der Solanaceaengewächse in Abhängigkeit von physikalischen Elicitoren während der Produktion und nach der Ernte.
ISBN 978-3-86247-066-2.

Band 15: Charles Adarkwah (2010)
Integrated management of the stored-product pest insects *Corcyra cephalonica, Cadra cautella, Sitophilus zeamais* and *Tribolium castaneum* by use of the parasitic wasps *Habrobracon hebetor, Venturia canescens, Lariophagus distinguendus* and neem seed oil.
ISBN 978-3-86247-077-8.

Band 16: Christoph von Studzinski (2010)
Angewandte Methoden der xenovegetativen Vermehrung.
ISBN 978-3-86247-088-4.

Band 17: Tanja Mucha-Pelzer (2011)
Amorphe Silikate – Möglichkeiten des Einsatzes im Gartenbau zur physikalischen Schädlingsbekämpfung.
ISBN 978- 3-86247-106-5.

Band 18: Diego Miranda (2011)
Effect of salt stress on physiological parameters of cape gooseberry, *Physalis peruviana* L.
ISBN 978- 3-86247-119-5

Band 19: Franziska Beran (2011)
Host preference and aggregation behavior of the striped flea beetle, *Phyllotreta striolata.*
ISBN 978- 3-86247-188-1

Band 20: Mohammed Abul Monjur Khan (2011)
Induced biochemical changes and gene expression in *Brassica oleracea* and *Arabidopsis thaliana* by drought stress and its consequences on resistance to aphids.
ISBN 978- 3-86247-203-1.

Band 21: Sandra Lerche (2012)
Untersuchungen zur Anwendung, Praxiseinführung und molekularen Identifizierung von Stamm V24 des entomopathogenen Pilzes *Lecanicillium muscarium* (Petch) Zare & W. Gams.
ISBN 978- 3-86247-248-2.

Band 22: Carsten Richter (2012)
Entwicklung und Überprüfung eines gasdichten Küvettensystems für Experimente unter hochgradig kontrollierten Bedingungen mit Gaswechselmessungen.
ISBN 978- 3-86247-271-0.

Band 23: Aksana Grineva (2012)
Influence of the two stored grain pest insects *Sitophilus granarius* and *Oryzaephilus surinamensis* on temperature, relative humidity, moisture content, and mould growth in stored triticale.
ISBN 978- 3-86247-279-6.

Band 24: Carmen Büttner & Christian Ulrichs (2012)
Aktuelle Themen in Landwirtschaft und Gartenbau am Beispiel von Südtirol.
ISBN 978- 3-86247-279-6.

Band 25: Juliane Langer (2012)
Molecular and epidemiological characterisation of Cherry leaf roll virus (CLRV).
ISBN 978- 3-86247-279-6.

Band 26: Franziska Rohr-Doucet (2012)
AOP-Variabilität in *Arabidopsis thaliana*-Kreuzungslinien – Auswirkungen auf die Resistenz gegenüber verschieden spezialisierten Lepidopteren-Arten.
ISBN 978- 3-86247-329-8.

Band 27: Vanessa Hörmann (2012)
Lignin als biologische Barriere gegen Schimmelpize in Innenräumen.
ISBN 978- 3-86247-330-4.

Band 28: Jacqueline Kurth (2013)
Auswirkungen verschiedener Düngerzusammensetzungen auf den Ertrag bei Schnittrosen unter Berücksichtigung des Anbauverfahrens.
ISBN 978- 3-86247-336-6.

Band 29: Juliane Langer, Carmen Büttner & Christian Ulrichs (2014)
Kolumbien – klimatische und politische Voraussetzungen für eine landwirtschaftliche Produktion.
ISBN 978- 3-86247-430-1.

Band 30: Heike Luisa Dieckmann (2014)
Detection of the European mountain ash ringspot associated virus (EMARaV) in Sorbus aucuparia L. in several European contries.
ISBN 978- 3-86247-441-7.

Band 31: Rima Marion Baag (2014)
Analyse von trans-Resveratrol in historischen Rebsorten der Weinanbaugebiete Sachsen und Saale-Unstrut.
ISBN 978- 3-86247-488-2.

Band 32: Ayesha Rahmann (2014)
Study of the protective effects of nano-structured silica and plant derived biomolecules on nuclear polyhedrosis virus affected silkworm larvae at the behavioral and molecular level.
ISBN 978- 3-86247-495-0.

Band 33: Bettina Gramberg (2015)
Weiterentwicklung eines elektrochemischen Biosensors zum Nachweis von Pflanzenviren und Insektiziden.
ISBN 978- 3-86247-512-4.

Band 34: Wilhelm van Husen (2015)
Artspezifische Aufnahme und Verteilung von Cadmium bei indigenen afrikanischen Gemüsearten und daraus abzuleitende Ernährungsempfehlungen.
ISBN 978- 3-86247-523-0.

Band 35: Jenny Roßbach (2015)
European mountain ash ringspot-associated viras (EMARaV): diversity and geographic distribution in Europe.
ISBN 978- 3-86247-547-6.

Band 36: Silke Steinmöller (2015)
Risikominderung der Verbreitung von Quarantäneschadorganismen der Kartoffel durch hygienisierende Maßnahmen.
ISBN 978- 3-86247-550-6.

Band 37: Christin Siewert (2016)
Genomic and functional analysis of species within the Acholeplasmataceae – Phytoplasmas and Acholeplasmas.
ISBN 978- 3-86247-579-7.

Band 38: Angela Köhler (2016)
Untersuchungen zur Phenolglycosidkonzentration ausgewählter intra- und interspezifischer Kreuzungen salicinreicher Biomasseweiden.
ISBN 978- 3-86247-581-0.

Band 39: Nicolas Meyer (2016)
Vergleichende ökophysiologische Untersuchung verschiedener Baumarten zur Verwendung als Straßenbegleitgrün in Berlin.
ISBN 978- 3-86247-586-5.

Band 40: Stefanie Schläger (2017)
Identification of variation within sex pheromone blends of various Maruca vitrata populations for refining pheromone lures and traps in Asia.
ISBN 978- 3-7369-9570-3.

Band 41: Elisha Otieno Gogo (2017)
Pre- and postharvest treatments for the quality assurance of African indigenous leafy vegetables.
ISBN 978-3-7369-9650-2.

Band 42: Luise Dierker (2017)
Interaktion des RNA2-kodierten Transportproteins (MP) des *Cherry leaf roll virus* (CLRV) mit dem viralen Hüllprotein (CP) und pflanzlichen Wirtsfaktoren
ISBN 978-3-7369-9670-0.

Band 43: Nadja Förster (2017)
Antikarzinogenes Potential ausgewählter Glucosinolate von *Moringa oleifera*
ISBN 978-3-7369-9704-2.

Band 44: Steffen Pallarz (2018)
Data driven classification of host-plant response (virus-plant)
ISBN 978-3-7369-9731-8.

Band 45: Vanessa Hörmann (2018)
Biofiltration of indoor pollutants by ornamental plants
ISBN 978-3-7369-9815-5.

Band 46: Allan Ndua Mweke (2018)
Development of entomopathogenic fungi as biopesticides for the management of Cowpea Aphid, *Aphis craccivora* Koch
ISBN 978-3-7369-9908-4.

Band 47: Isabella Linda Bisutti (2019)
Biological agents formulation and mode of application against strawberry diseases
ISBN 978-3-7369-7032-8.

Band 48: Judith Henze (2019)
Innovation in Agriculture: The Potential, Challenges and Adoption and Diffusion of Aquaponics and Agricultural Mobile Phone Application in Kenya
ISBN 978-3-7369-7133-2.

Band 49: Maliha Gul Aftab (2021)
Chemical ecology of Cabbage White (Pieris sp.)
ISBN 978-3-7369-7498-2.

www.ingramcontent.com/pod-product-compliance
Ingram Content Group UK Ltd.
Pitfield, Milton Keynes, MK11 3LW, UK
UKHW022001190726
13853UKWH00004B/1664

9 783736 975439